THE COMPLETE IDIOT'S GUIDE® TO

Electrical Repair

by Terry Meany

alpha
books

Macmillan USA, Inc.
201 West 103rd Street
Indianapolis, IN 46290

A Pearson Education Company

Publisher
Marie Butler-Knight

Product Manager
Phil Kitchel

Associate Managing Editor
Cari Luna

Acquisitions Editor
Randy Ladenheim-Gil

Development Editor
Alexander Goldman

Production Editor
Christy Wagner

Copy Editor
Amy Lepore

Illustrator
Jody Schaeffer

Cover Designers
Mike Freeland
Kevin Spear

Book Designers
Scott Cook and Amy Adams of DesignLab

Indexer
Brad Herriman

Layout/Proofreading
Fran Blauw
Mary Hunt
Liz Johnston

Contents at a Glance

Contents

Foreword

Some years back (in high school as a matter of fact) I was involved with the technical side of the drama department. On one particular occasion, while we were hanging and testing lighting fixtures, I happened to look over to one of my classmates, who was grasping a fixture and the steel railing unusually tightly, and whose hair was defying certain laws of gravity.

He was the recipient of a few spare volts from a lighting fixture that was not properly grounded. It was probably at that moment that I developed a great respect for electricity. (My classmate suffered no long-term damage—though he did become an actor ...)

After many years working in theatre and architecture, I have seen bizarre electrical work—some by homeowners, some by electricians. What separates good electrical work from the bizarre is the cleanliness of the job. I know one electrician who, if he nicks a piece of cable anywhere along its run, will pull it out and start over, no matter how long the run. It is that attention to detail, that striving for perfection, that makes him such a good electrician. He understands that there is little room for error. You should follow similar standards.

Electricity has simple rules—this manual gives you a good insight into those rules. Please consider reading the entire book before you jump into a single project. There are so many good tips spread throughout the manual. Two most important things: 1) Turn off the power before you do any work. Resetting the clocks is much easier than resetting your cranium. 2) Know your limitations. If you have any doubts, call a licensed electrician.

A good plan is also helpful for your projects. All electricians work from blueprints. Drawings help to organize the whole project. A good electrician will have a number of drawings and will outline exactly how wires will be pulled throughout the project. This advance work can save hours of frustration and repair time spent on extraneous holes in walls.

And please don't underestimate the power of new lighting. Chapters 13 and 20 provide examples of ways to beautify your home. Just by changing your standard household screw-in light bulb to a directional PAR lamp you can change the entire appearance of your environment. Additional lighting on work surfaces can improve vision and make tasks easier.

The Complete Idiot's Guide to Electrical Repair is like no other reference manual. I teach Lighting Design at New York's Fashion Institute of Technology, and in the past when it came to teaching about electricity, no book existed that so clearly and thoroughly covered electricity and wiring. I am pleased to put *The Complete Idiot's Guide to Electrical Repair* on my list of textbooks. This is not just a manual for beginners—handymen and other advanced homeowners will find invaluable information and tips to make wiring faster, easier, and less expensive.

I wish you all good fortune on your future projects, and don't forget to secure the ground wire.

Matthew Tirschwell
President
Tirschwell & Co., Inc.
Architectural Lighting Design

Introduction

Electrical wiring, fixtures, and appliances have been part of our homes for almost a century, and they've been a wonder, unless your system is almost a century old! Then you have to wonder if it's safe, let alone satisfactory to meet the demands of a modern lifestyle. Even if you have a newer system, you may still want to make additions to it and extend its capabilities. In principle, this is just another remodeling job, but we treat wiring differently. Adding a circuit isn't the same as adding a cabinet.

A poorly planned or installed cabinet won't shock you or create a fire hazard. Nor does it require a permit and an inspection. You can hang it crooked or hang it over a water pipe, and it will still do its job. Electrical work isn't so easy, but it isn't incomprehensibly difficult, either.

Many of us have little understanding of our electrical systems, or electricity itself for that matter, so we call electricians when we can't figure out why the lights keep going out or when we want to add a receptacle to a bedroom. Even in an age of supermoms and multitasking dads, we can't know how to do everything, but does electrical wiring need to be all that daunting? No, it does not, as you'll see by the time you've finished reading this book.

Big jobs, like installing a new electrical service, are usually best left to professional electricians, but anyone with a few tools, some elementary math skills, and a free weekend can add a circuit or replace old light fixtures. Wiring is a relentlessly logical process (well, that and a lot of drilling and pulling). The rules are clearly spelled out and easy to follow. You can put away your unwarranted fears about electricity—but not your precautions—and safely do much of your own work.

The chapters that follow will give you a better understanding of just what electricity is and how wiring systems work. We'll walk through the steps for everything from replacing a switch to wiring a bathroom. As you read, the mystery will slowly wear off as you start planning more lights, receptacles, and upgrades. You can even automate your house and set it up like one of the bad guy's fortresses in a James Bond movie.

Like any remodeling project, upgrading or adding on to your electrical system will require planning and a budget, at least for the bigger jobs. Large jobs, such as rewiring the bulk of your house, should be broken down into smaller jobs so they're less overwhelming. If you try to do too much at once, it's easy to find yourself with a jumble of wires, all of your circuits disconnected, and the end of the day approaching. Remember, you're learning some new skills. You won't become a master electrician overnight.

Electrical work brings some secondary tasks along with it. In some cases, you'll have to open up your walls and ceilings, and this means patching those holes later. Patching is usually followed by painting. It's tempting to let this go since it's surprisingly easy to let three or four years pass by looking at partially patched and unpainted walls. Be sure to give yourself enough time to complete the entire job.

Finally, remember why you're doing these projects: to make your home more comfortable, up-to-date, and safe.

How to Use This Book

Working on your house can be like raising children: Every day is an adventure. You want as few adventures as possible when you work around electrical wiring, though. In fact, one good-size adventure could be your last if you manage to shock yourself in a big way. Unlike other remodeling projects, electrical work is less broadly disruptive (you're not tearing out entire walls, for instance—at least, you'd better not be), which is a big plus.

This book is set up to give you a broad overview of electricity and systems first and follow up with actual projects, starting with the simplest. It's not an apprenticeship, but you'll have enough information to evaluate your system and make intelligent decisions about its condition and any need to upgrade. And you'll be a little more savvy when hiring an electrician.

Your work must always follow your local codes. Beyond that, you can add circuits and gadgets to your heart's content.

How this book is organized:

Part 1, "The Basics: Out of the Dark Ages": Before you do any electrical project, you need to know how your system works, where all those wires go, and what a service panel does. Snoop around your panel or fuse box and check out all of your electrical devices so you'll know what you're dealing with.

Part 2, "Safety, Tools, and Contractors": Many construction companies claim that safety is their first concern, and it should be yours, too, especially when you work around electricity. The right tools are always a must, whether you buy, borrow, or rent them. A few words about contractors are included here, too, should you decide to hire the work out.

Part 3, "Components and Simple Repairs": You have to start somewhere, so I'll start with defining switches, receptacles, and fixtures and then discuss how to repair and replace them. Troubleshooting skills will make some repairs easier and faster.

Part 4, "Power Hungry": Part 4 deals with the big jobs: a new service panel and running circuits to kitchens, bathrooms, and outdoors. If you don't have gas or oil, you should read about electric heat (air conditioning, too).

Part 5, "Refinements": Once you've taken care of the basics, you'll want to do more. Workrooms, low-voltage wiring, and security systems all have their say in this part. And who doesn't need a doorbell? Finally, a few thoughts on conserving electricity.

Acknowledgments

Few books are solo efforts, and this one is no exception. I'd like to credit everyone whose generous efforts and contributions helped bring this manuscript together.

I'd like to thank my technical editor, Don Harper, who corrected me on more than a few occasions. When I least expected it, a fax would come over the line with the relevant electrical code and his notations on it.

Images are everything in a how-to book, and I am grateful for the artwork provided by Pamela Winikoff at Leviton; Raymond Venzon of Makita USA.; Mike Mangan of MKM Communications; Joyce Simon at Western Forge (for Craftsman Tools); Pat Gengler (Milwaukee Electric Tool Corporation); Paige Malouche, Marketing Services Manager at Progress Lighting; the Wiremold Company; Saverio Manciniof Mintz & Hoke, Inc.; and Tom Monahan. Kibby Bowen, along with her husband, Brock, provided the black-and-white photography.

On the writing side, Randy Ladenheim-Gil at Macmillan and Alex Goldman handled the editing and have my thanks for doing so.

Christy Wagner at Macmillan put it all together.

Finally, my gratitude to my agent, Andre Abecassis of the Ann Elmo Agency, who keeps finding me such interesting assignments.

Special Thanks to the Technical Reviewer

The Complete Idiot's Guide to Electrical Repair was reviewed by an expert who double-checked the accuracy of what you'll learn here, to help us ensure that this book gives you everything you need to know about home electrical repair. Special thanks are extended to Don Harper.

Don Harper is a licensed Washington State electrical contractor and holds both an electrical administrator certificate and an electrical journeyman card. He is a graduate of the Construction Institute Trades Council and has taught first-year electrical classes there for seven years. His company, Harper Electrical, does both new and remodeled residential wiring as well as installations for high-tech communication and software companies.

Trademarks

All terms mentioned in this book that are known to be or are suspected of being trademarks or service marks have been appropriately capitalized. Alpha Books and Macmillan USA, Inc. cannot attest to the accuracy of this information. Use of a term in this book should not be regarded as affecting the validity of any trademark or service mark.

Part 1

The Basics: Out of the Dark Ages

In many ways, life was much simpler before the advent of electricity. People slept longer—after all, there wasn't much else to do when it got dark—and worked fewer hours for this same reason. Candles and gaslights just didn't cut it when it came to providing safe, well-lit working and living spaces.

In addition to lighting the way, electricity powers just about everything you touch and use. You should be able to enjoy all the benefits of a wired home—lights, receptacles, and the toys of civilization—wherever you want them. This is a doable goal regardless of the age of your house or its wiring. With some basic knowledge and understanding of your electrical system, you can surround yourself with power where you want it and have conveniences at your fingertips.

Before you start snipping away at your old knob-and-tube wiring, read through these first few chapters and get the basics. You'll find out how electricity flows from your local utility to your espresso maker in a safe, predictable manner and how you can keep it that way. All the wires running through your walls want to live an orderly life and have no interest in the anarchy of bad wiring jobs (which are not an uncommon problem in old homes, unfortunately). You don't want a future homeowner uncovering your work and wondering, "How could anyone wire like this?" It won't happen after you've gotten these chapters under your tool belt.

Chapter 1

Fear of Frying

In This Chapter

➤ The logic behind your electrical system

➤ Getting the job done

➤ A brief inspection of your wiring

➤ Fuses and circuit breakers

➤ Running power where you want it

I once had a client who was installing some light fixtures in his Seattle home. While he was working, his mother called from New York. When told by her daughter-in-law what her son was doing, she screamed, "You tell him to get down. Doesn't he remember what happened to Mr. Schneider down the street? He got electrocuted doing such things. What is he thinking, this son of mine?"

Mr. Schneider, it seems, didn't know very much about electricity or his home's wiring. Electricity isn't some kind of barely contained liquid fire inside your wiring just waiting to strike and burn innocent victims. It's a civilizing force in our lives that we won't live without. Even when we go camping, we often take battery-powered gadgets so we can rough it in comfort.

This chapter will show you that your house's wiring, if done correctly and legally, is a nice, logical system that should be respected, not feared. You'll get a better feel for the work involved in upgrading or altering your system. You also will start to think about changes and improvements you might not have considered previously. Think of this as a bare-bones introduction to get you thinking about your electrical system and how to upgrade it.

You'll also learn to do your work safely without worrying your mother too much.

A Wired World

We take electricity so much for granted that it's hard to believe many rural parts of this country lacked electrification until the 1930s. Now we have it in every room of the house, the garage, the basement, and even outdoors. Chapter 2, "What Is Electricity Anyway?" will get into the science of electricity. As a homeowner, what do you need to know before you start working on your wiring? What should you be looking for?

An electrical system is composed of a variety of parts, from those as large as a dam or another power generator to others as small as the wiring attached to your doorbell. The power coming into your house is much too powerful to use safely at full strength. Instead, it's broken down into smaller units through a system of circuits with breakers or fuses and different-size wires. Every component along the way has a role to play. Unlike income taxes, this is a very logical system.

Linear Logic

Left to its own devices, electricity wouldn't be much good to us because it requires some discipline to be useful. This discipline, in the form of electrical current, corrals the charged electrons that make electricity and directs them so they can power our lights, computers, and electric apple peelers. Your local utility company's generators produce the electricity and then "pipe" it to your home through wires and trans-formers. The only time this is of any great interest to you is when there's a disruption in the distribution system that results in your power going off and your digital clocks reverting to that annoying, flashing 12:00 signal when the power comes on again.

Electrical Elaboration

A utility company's circuits can get overloaded just as circuits can overload in our own homes. Too much demand for power to run fans and air conditioners during hot spells, for instance, can cause a loss of power for entire neighborhoods. Trees are another culprit. All it takes is one branch falling across some power lines to disrupt electrical service to any-one depending on those lines. For this reason, power companies maintain ongoing tree-trimming programs, which can be a difficult task in large rural areas. When a utility can foresee excessive, short-term demand, it might selectively shut down power if it can't pur-chase additional power from another utility.

Once the power lines enter your house, your interest naturally perks up. Here, the comfort and safety of you and your family are your number-one concerns.

Follow the Electrical Code

The installation of electrical systems in the United States is subject to local building codes. As a rule, these requirements are based on the National Electrical Code (NEC). (Canadians use the Canadian Electrical Code, or CEC.) The NEC carries no enforcement power and is written as an advisory document only, but for all intents and purposes, this is the main set of rules on which local codes are based.

The NEC is the guiding authority for electricians and is not exactly bedtime reading for the rest of us. Local codes might be more stringent in some areas. As a homeowner or an electrician, you have to be aware of any specific rules that your local codes might impose.

Electrical codes spell out, among other things ...

➤ Lighting requirements

➤ Receptacles needed per square foot of living space

➤ How the system should be grounded

➤ Circuit sizes

➤ Required wire gauge or size per individual circuit

➤ Special stipulations for kitchens, bathrooms, hot tubs, pools, fountains, and outdoors

Positively Shocking

The National Electrical Code (NEC) is designed strictly as a safety measure to protect you and your property. It is not meant to be an instruction manual for amateur electricians or to be used as a design specification for your home or business. The NEC covers most, but not all, electrical installations.

Codes are like personal relationships: Everything can be going along just fine until there's a misunderstanding or a misinterpretation of something someone has said. Then all interested parties have a problem. Electrical inspectors and electricians, both professional and do-it-yourselfers, sometimes have different interpretations of the code. For this reason, you want to be absolutely sure your work is done in the most straightforward manner possible, even if it means a little more expense or work on your part. After all, regardless of your interpretation, it's the inspector who makes the final ruling. The authority having jurisdiction of the code will have the responsibility for making interpretations of the rules (Article 90-4). Leave literary license to wayward authors.

Safety Rules, Mr./Ms. Homeowner

It has been suggested that early electricians at the turn of the century were a paranoid lot. This was a new, untested medium that was replacing familiar gas lighting. These electricians weren't interested in developing reputations as de facto arsonists. Wiring at the time was pretty simple to begin with, usually just lighting circuits, one receptacle per average-size room, and a very small service or fuse box. Electricians used lead solder followed by tape to join wires and do their work safely.

Your dealings with electricity should be equally safe, whether you're installing a new circuit or screwing in a light bulb. Electricity always is seeking an easy way to travel. Sticking your fingers, screwdrivers, or car keys into light sockets or receptacles provides these charged particles with an alternative path to moving along a wire. An improperly grounded toaster can cook more than your bagels. We'll cover the basic safety rules in Chapter 7, "Caution Signs and Safety Concerns." For now, you'll need to keep a few rules in mind when dealing with your electrical system:

➤ Don't handle anything electrical if you're wet or are standing on a wet surface.

➤ Never overload a circuit beyond its capacity.

➤ Extension cords are for temporary use only.

➤ Never start an electrical repair or addition until you're sure how to do the job correctly and the power is shut off.

➤ When a problem is beyond your expertise, call a licensed electrician.

Bright Idea

If you have to change a fuse or check a circuit breaker in an area where the floor might be damp, lay down a piece of plywood first. Standing on this will keep you on dry ground, which is less hazardous than damp concrete. You also should wear dry, rubber-soled shoes and leave one hand in your pocket to keep from inadvertently becoming a pathway for the electrical current.

Mutual Respect

Franklin D. Roosevelt said that the only thing we have to fear is fear itself. He obviously never dealt with the IRS. We can include electricity as one thing we don't have to fear, but we do need to respect it. You and your electrical system will get along just fine as long as you don't demand more of it than it's designed to provide. Most problems with electricity result from poor workmanship, code violations, and user abuse. Old systems were designed to power far fewer toys and gadgets than we have today. Trying to run three or four small kitchen appliances out of one receptacle, rather than running a new circuit, is just asking for trouble.

Do It Yourself or Hire It Out?

Electricians are one of the elite—and expensive—building trades. They are trained and tested to become licensed (a must when you're hiring). They most likely can do a large job faster than you can. As with any trade, *electricians* come equipped with the tools and knowledge that you are now just beginning to acquire. This doesn't mean you aren't up to the challenge—for most jobs, you will be. Once you understand how to run new circuits, replace lights, and upgrade old wiring, you'll be able to do your own electrical work in a professional manner.

In addition to having a working knowledge of the code requirements and knowing how to install your wiring and fixtures, just what does this work involve? The following sections explain this in more detail.

Drilling and Pulling

The physical act of wiring is largely a matter of getting power from point A to point B in a manner approved by the code. Point A might be your main service panel (where the power enters your house), or it might be a receptacle on an adjoining wall. Either way, you have to figure out the best route to run your wire so A and B can be connected.

How do you define the best way? That depends on your circumstances:

➤ Are your walls and ceilings open with the studs and joist exposed?

➤ Do you have to work around old plaster and lath or newer drywall?

➤ Is there basement, attic, or crawl-space access?

Much of an electrician's time is spent drilling holes in wall studs and floor joist and pulling electrical cable from one fixture or receptacle to another. This work is tougher in a finished house, especially one with old plaster walls or limited access from either a basement or attic crawl space. This is time-consuming work, and its cost can be difficult to estimate. In my opinion, these are perfect jobs for homeowners who can take their time drilling and "fishing" wires even if they don't want to do the final connections or fixture installations. A couple of weekends or evenings with a commercial-quality drill and a roll of electrical cable can greatly reduce the time an electrician spends in your house—and can greatly reduce your costs.

Ask an Electrician

In terms of training and expertise, an **electrician** starts out as an **apprentice** before moving up to **certified journeyman** status. With additional training and testing, he or she can become a **master electrician.**

Bright Idea

It's always easier to have two people feeding and pulling wires between floors, even if you can do the work alone. Kids can get in on the work, too. This gives them a sense of accomplishment as well as some basic knowledge of how wiring works. This is a great skill to have when they're older and are wiring their own homes.

Neatness Counts

I cannot emphasize enough the need for clean, neat, and accurate work when doing your own electrical jobs. Inspectors aren't fond of homeowners doing their own wiring, and they probably will scrutinize your work more than the work of an electrician. Chalk it up to one more example of life being unfair, or see it as motivation to do the best work possible. (How's that for making lemonade out of lemons?)

A new electrical service that's been done well is a beautiful exercise in symmetry. All the wires entering the service panel are installed at neat right angles without any excess length. Wires running along exposed basement floor joist are taut, stapled, and secured. The point of the staple is to gently hold the cable in place. It is very easy to damage the outer sheath of NMB (nonmetallic) cable if you aggressively pound staples against it.

Cable inside receptacle and switch boxes is cut clean and is folded in and out of the way at the back of the boxes. These are not inordinate standards but the ones an inspector expects to see. You should expect them, too, whether you do your own work or hire it out.

Can you get these results as a novice? Of course you can! It will take you longer than a trained electrician, but so would just about any work that's new to you. That's why you bought this book. This text—and a few good tools (see Chapter 8, "Call Me Sparky")—will see you through most electrical jobs with inspector-pleasing results.

Electrical Elaboration

A good electrical inspector will work with you on a project and will inform you of possible missteps that might be in the making. On my first commercial job, the inspector didn't say a word to the electrician about the way he was routing his cable between floors until the job was almost finished. At that point, she told him it wasn't correct and would have to be redone. They disagreed about how to interpret the code, but nevertheless, she should have brought up her concerns earlier. He didn't have to reroute, but he did have to change some panel boxes, which could have been avoided had they both communicated more clearly.

Simple Projects First

Before you go yanking your old fuse box out, convinced that you can replace it before dinnertime with new circuit breakers, look for a small job to do first. Most older homes have at least one receptacle or switch that needs replacing. There are other jobs to consider as well such as …

➤ Adding extra garage lights.

➤ Running a dedicated circuit for your office computer.

➤ Installing a bathroom fan.

➤ Adding lights to your backyard.

These are good jobs for practicing your evolving electrical skills without causing too much disruption around your house. They all involve applying for a *permit*, scheduling an inspection, calculating an electrical load, running wire from a power source to a fixture, installing the fixture, and making the final connections of wire, fixture, and power source. Each of these jobs is a microcosm of a larger project such as re-wiring your entire house, and each is a good confidence booster. You can even take snapshots of your work to carry around in your wallet, but be prepared for some strange looks from your friends when you pull them out for showing.

System Checkup

By now, you're probably getting some ideas for the kinds of projects you might consider doing, but what do you really need to do? What shape is your electrical system in now? The newer the house, the more likely there is less code work to do. That is, you shouldn't have to correct any existing wiring if it's original to the house. This isn't an absolute rule, however! Sometimes an inspector misses something or an owner does some work that isn't up to code.

Older houses are more problematic. It's common to find a jumble of add-ons and questionable work in an old home. Even a cursory inspection will give you some idea of electrical improvements you might consider making.

Ask an Electrician

Any work that extends an existing electrical system by adding a circuit or a fixture usually requires a **permit** and an inspection. Any work that simply replaces an existing fixture, such as a light or a receptacle, usually doesn't. Always check with your local building department to be sure.

Plugless in Seattle

One of the biggest drawbacks to old wiring systems is a lack of receptacles or outlets. Remember, our parents' and grandparents' generations had far fewer voltage-eating consumer trinkets and entertainment devices than we have today. Current code calls for …

➤ A receptacle to be installed so a six-foot cord can be plugged in anywhere along a wall in general living areas.

➤ Special ground-fault circuit interrupter (GFCI) outlets to be installed in kitchens, bathrooms, near any sinks, and outdoors.

➤ Special considerations for floor-mounted outlets.

Could you use some additional receptacles? Is your bathroom receptacle up to code with a GFCI? Look around your house to see if you could use some additional receptacles. Also make sure your bathroom receptacle has a GFCI, as code requires.

Let There Be Light

Parents and teachers of a certain generation regularly reprimanded children to do their reading in "decent" light, warning that they could "ruin" their eyes by using dim lights. Whether you believe this to be a medical fact or not (I've heard it both ways), why not give yourself as much light as possible when you read or do other close work? Adding lighting where you want it is one of the great benefits of electrical wiring.

Positively Shocking

When replacing lights, don't assume you can install a light with higher wattage. The circuit might not support the additional power demand. You always should confirm the total demand by other lights, receptacles, or appliances before changing an existing fixture.

Lighting fulfills other purposes besides purely practical ones. It can set a mood, spark romance, and ward off ne'er-do-wells lurking outside on a dark night. Yard lights invite summertime parties and welcome us home in the winter. Adding additional lighting is a more complicated job than simply adding a receptacle, but it certainly is within the scope for a homeowner to do.

Hot Spots

Any receptacle or switch that is hot to the touch is an overloaded circuit. This is a circuit that is drawing more current than it's designed to draw. If you have any hot spots, you must attend to them immediately. (You can start by pulling a few plugs or turning off the lights.) A shortage of receptacles or lights is an inconvenience; an overloaded circuit is a danger that should not be ignored.

Special Needs

Every home and homeowner is different. What might have been a perfectly acceptable electrical system for a previous owner might be woefully deficient for you. Maybe you want to install a small baseboard heater in a bathroom located a long way from the furnace. Your photography hobby might demand a darkroom. As an antique car restorer, you can't wait to set up a paint booth in the garage complete with industrial heating units for that baked-on finish guaranteed to win a trophy or two.

New houses often are constructed with the minimum number of code-required receptacles and lighting. Exceptions are made with kitchen and bathroom lights; these are high-profile areas that help sell houses, so builders make them brighter and more appealing with better lighting. Old houses often have a real hodgepodge of wiring that you'll probably want to upgrade and expand. One of the reasons you're reading this book is to custom design and improve your electrical system to suit your needs, not those of a builder or a previous owner.

Confused About Fuses?

Every fully electrified house has either a fuse box or a main panel box with circuit breakers. This is the distribution center for the power coming into your house. Without them, you would have one whopping current running through your walls that would burn out just about any appliance you tried to run on it.

Fuses were used until approximately 1950, when circuit breakers became the standard installation for new construction. The fuses most of us are familiar with are the round, screw-in glass types with a visible alloy strip inside the glass. These are called plug fuses. Cartridge fuses, which have a cylindrical shape, are the other common type of fuse.

If the current running across a plug fuse's alloy strip exceeds the amperage of the fuse, the strip will melt, thus stopping the flow of electricity. There is nothing inherently wrong with a system using fuses, but they are dated and inconvenient. If you don't have any spares around when one "blows"—you should always replace a fuse with one of the same amperage—you're out of luck. The other problem with plug fuses is that a fuse with an amperage setting of 15, 20, 25, or 30 can be installed as a replacement for a burnt-out fuse even if the original size should have been 15 amps. Even though it is physically possible to install the wrong fuse, doing so could overload a circuit and might even start a fire in your home. To prevent this, the installation of an "S" type adapter will limit the maximum fuse size to 20 amps.

Circuit breakers, the modern standard for homes, are an improvement over fuses, as you'll see in the next section.

Bright Idea

If you're uncomfortable putting your hand into a fuse box to replace a burned-out fuse, you can buy a tool to help with the job. Electrical-supply companies sell plastic fuse pullers specifically made for gripping fuses and removing them. The plastic will not conduct electricity, so there's no danger of receiving a shock through them.

Electrical Elaboration

Although circuit breakers are the standard equipment for circuit protection in your home, fuses are still used in many other applications. Fuses with ratings as high as 10,000 amps and 136,000 volts are used in marine, automotive, telecommunications, and computer applications. Fuses provide circuit protection in motors, transformers, and an array of delicate electronic equipment.

Circuit Breakers

A circuit breaker serves the same function as a fuse, but it's a more complicated device. It also is reusable. When a current that exceeds the breaker's capacity or rating passes through it, a pair of metal contacts is broken and remain so until the breaker is reset. A breaker can be reset an almost indefinite number of times, although repeated tripping is a sign of an electrical problem or overload. Any time a breaker trips or a fuse burns out, you must find the source of the problem before you reset the breaker or replace the fuse. Sometimes it's only a single-occurrence problem such as running too many appliances at once. If you can't find an apparent cause in your use of the circuit, you probably have a short in the system that must be addressed (see Chapter 16, "Trouble, Troubleshooting, and Safety").

More Power to the People

Modern electrical systems give us access to plenty of safe, dependable power. Around the turn of the century, it was a big deal to have a 60-amp service. Now, 200-amp services are common in many houses, and some larger homes are even getting 400 amps of power. We are dependent on electricity for our safety and well-being. One purpose of this book is to help you put it to the best use possible in your home.

Any electrical system can be improved and adapted to your individual needs and specifications:

➤ A larger service of greater amperage can be installed.

➤ New circuits can be added.

➤ Existing circuits sometimes can be extended.

➤ Lights can be added anywhere there is a need for them.

➤ Additional wiring can facilitate modern contrivances from garage door openers to barbecue rotisseries.

As you read on, you'll learn how to perform these electrical chores by yourself or how to evaluate your needs and discuss them intelligently with an electrician. Either way, you'll have power at your fingertips throughout your home.

The Least You Need to Know

➤ Electrical systems are logical, precise, and guided by local electrical codes.

➤ A do-it-yourselfer can safely do many electrical projects around the house, but he or she should start with simple jobs first.

➤ A simple walk-through of your house and yard will give you some improvement and upgrade ideas for your electrical system.

➤ Take the time to design your electrical system or upgrades to suit your needs, lifestyle, and sense of convenience.

Chapter 2

What Is Electricity Anyway?

In This Chapter

➤ How alternating current works

➤ Staying grounded for safety

➤ Knowing volts, watts, and amps

➤ Wire size matters

We use and depend on electricity every single day. All we usually know about it is that it's buried inside our walls, it runs our lights and VCRs, and we're billed for it every month or so. Terms such as kilowatt hours, amperage, volts, and current are Greek to most of us. This is probably appropriate because the Greeks first described static electricity about 2,500 years ago. It was discovered that amber would accumulate a negative charge of static electricity when rubbed with sheep's wool. Not known for a great sense of comedy, this probably became quite the party trick at Greek get-togethers. The word "electricity" has its root in the term *electrum,* which is Latin for "amber."

Understanding electricity is like understanding cooking: Once you know a bit about sautéing, cooking temperatures, seasonings, and how to make a decent pie crust, you can muddle through meal preparation and come up with more-than-edible results. If you know how electricity is produced and can toss around some vocabulary words, such as alternating current and resistance, you'll be more comfortable with your electrical work. A task makes more sense when you understand its inner workings.

This chapter isn't going to give you enough information to challenge a Ph.D. in electrical engineering to a trivia contest at your local *Jeopardy* theme bar. You will, however, develop a working knowledge of electricity basics and how they apply to your own electrical system.

Go with the Flow

Think back to your high school physics classes and all those diagrams of atoms with electrons spinning around a nucleus. (They're the drawings that looked like really small solar systems.) Basically, electrons spin around because the protons in the atom's nucleus carry a positive charge (+) that repels the electrons' negative charge (–). If enough of the electrons decide to move on, preferably in a more or less uniform stream, we end up with usable electricity.

Electricity comes in several flavors, but the two we're most familiar with are …

➤ Static electricity, in which the electric charges are stationary.

➤ Dynamic electricity, in which the electric charges are moving in a current.

When you were younger, the main value of static electricity was using it to shock unsuspecting siblings and cousins after you had walked across a carpet. If you didn't do this when you were a kid, you can always try it at your next holiday dinner. Cats also are good targets, but their revenge usually is a messier affair.

Why does the shock occur? Because some electrons like to travel, and they aren't the most stable subatomic particles. When you walk across a carpet (some are worse than others), you pick up some of these hitchhiking electrons while leaving some of your own positive charges. They have to go somewhere, and your sibling's finger or a doorknob makes a dandy conductor. If you touch a door frame, nothing happens because wood is a good insulator. That is, it does not allow electrons to easily move through it.

Static electricity is simply an imbalance of positive and negative charges. When you get zapped, you're just the accountant trying to balance these charges. One place you don't want to balance these charges, by the way, is with your computer, so you can either …

➤ Touch your metal desk chair before turning on your computer to get rid of any pesky electrons that could affect your computer.

➤ Apply anti-static spray periodically to your carpet so it will have a more positive charge and be less likely to give up its electrons.

Static electricity may be annoying, but dynamic electricity is another story altogether.

Staying Current

Electricity doesn't do us much good if a bunch of errant electrons constantly change orchestras from one conductor to another. We want our electrons to move in a reasonably orderly fashion so they can do our bidding when we turn on the lights. A flow of electricity is called a current, and it's carried into our homes through wiring from local electric utility companies. New electrical systems have the following three wires coming into your house:

➤ Two black or "hot" wires that carry the current to your service panel

➤ One bare neutral wire for carrying the current back to the power source and to ground

An electrical current has a couple of different options, depending on your application.

AC/DC

When Thomas Edison and his crew invented a reliable electric light bulb, he followed it up by developing the power systems to run it, rightfully envisioning a future world full of light bulbs. (We usually refer to these as "light bulbs," but "lamps" actually is the correct term. Bulbs are for planting.) Edison employed direct current (DC), which now is used in battery-operated gadgets in which the current flows from the negative terminal of the battery to the positive terminal. A battery is basically a container of chemicals whose electrochemical reactions produce excess electrons.

Our electrical systems use alternating current (AC), which was developed by Edison's contemporary, George Westinghouse, after he bought up patents from Nikola Tesla and William Stanley. Once again, someone with business sense trumped the scientific minds possessing the money-making ideas. It took Edison, the lampmeister, a few years to go along with this AC business, but he eventually told Westinghouse's son to let his dad know he was right.

A direct current just means that the electric current flows continuously in one direction and keeps going until it finds something to run such as a radio or a light bulb. An alternating current flows in one direction—say, to a receptacle—and then flows back in the opposite direction. You might be thinking, so what? When was the last time alternating current was discussed on late-night talk shows? Probably never. Alternating current, however, does have some useful, consumer-friendly features such as the following:

➤ Through a series of transformers, an AC can be increased or decreased in value. (The current can be made stronger or weaker.) This means that, instead of a zillion watts of power heading for your panel box, you'll get a reduced amount that you actually can use.

➤ Alternating current is efficiently transported over long-distance power lines.

➤ It's easy to convert from AC to DC, but it's expensive to go from DC to AC.

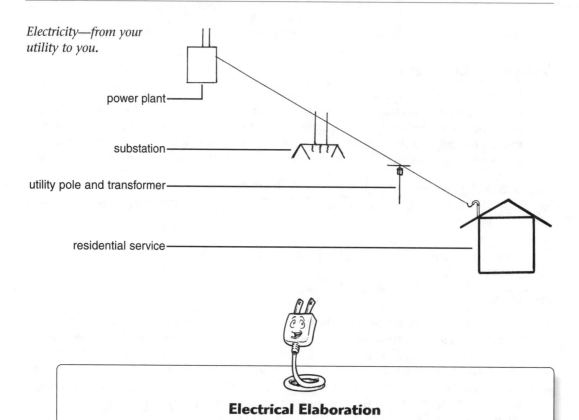

Electricity—from your utility to you.

power plant

substation

utility pole and transformer

residential service

Electrical Elaboration

George Westinghouse installed the world's first long-distance AC power lines in the unlikely area of the San Juan Mountains near Telluride, Colorado. Rapid expansion of gold mining had exhausted nearby timber supplies, the cheap fuel that ran the steam-powered machines the mines required. L. L. Nunn, a partner in the Gold King Mine, traveled to Pittsburgh and convinced a reluctant Westinghouse to build a generator and motor in the isolated mountain region despite its freezing temperatures and avalanche dangers. Nunn threw the switch on June 21, 1891, beginning a new era in electric power in the United States.

You're Grounded

Now that you know what kind of current you have in your house (and everywhere else), let's discuss another critical feature—grounding. Your entire electrical system, if it's up to current code, is grounded for your protection. This literally means that one wire of your electrical system leads back into the earth itself, where it will carry any errant current that could otherwise shock or electrocute you.

The earth ends up being a good electrical conductor and a convenient return path for electrons. In fact, the earth is used as a reference point for measuring the voltage in our electrical systems.

A ground wire can be attached to a ground rod that is deeply buried, or it can be a length of copper wire buried near your foundation's footings. A second physical ground is usually your cold-water supply pipe near your service panel.

Modern house wiring is color-coded so you won't confuse your hot, neutral, and ground wires with one another. This coding is standard everywhere—there is no room for artistic creativity here. The wire colors are …

➤ Black and red for hot wires

➤ White for neutral wires

➤ Bare (unsheathed) copper or green for ground wires

The black, red, white, and green colors refer to the plastic sheathing that contains the wires themselves. If you have an old two-wire system (see Chapter 5, "More Wall Talk"), you won't have a ground wire. An old knob-and-tube system (see Chapter 3, "History Lessons") sheaths both the hot and neutral wires in black, which isn't exactly user-friendly when you're trying to distinguish one wire from another.

It's important to understand the difference between the grounding wire and the neutral wire. The neutral white wire carries the electrical current back to the power source after it's passed through a *load* (a ceiling light, a fan, a stereo, and so on). That's the nature of an alternating current. The grounding wire, on the other hand, protects the entire system. The neutral wire is more correctly referred to as a *grounded conductor*. The bare or copper wire is a *grounding conductor*.

Ask an Electrician

The term **load** refers to any device that uses electricity. Your house is basically one large cumulative load, and lights, appliances, and air conditioning are individual loads.

Voltage Provides the Push

There are a lot of terms associated with electricity. Different words refer to a current's strength, the speed at which it travels, and the rate at which it's consumed. *Voltage* is a sometimes-misunderstood term that means "electomotive force" or, more simply, electrical "pressure." Voltage also is the difference in electrical potential between one end of a circuit and the other. In our electrical systems, voltage is measured against the earth, which is at zero potential. In other words, it all starts with the ground under your feet. Voltage gets the electrical ball rolling by giving a push to electric power from your utility's generator to your house or business.

Bright Idea

If you're working with electricity or with electrical tools outside your house, you can't avoid some contact with the ground. You can minimize any shocks by using a wood or fiberglass ladder. (Note: Electricians aren't allowed to use metal ladders.) Wearing boots or shoes with thick rubber soles also will provide insulation between you and the ground.

Positively Shocking

Europe and other parts of the world use a 220/240-volt electrical system. Before you can plug in your notebook PC, razor, or hair dryer, you have to use a converter to reduce the current to 120V. Converters are available at travel specialty stores and many overseas hotels.

Long-distance power lines carry huge voltages, from around 155,000 to 765,000 volts. If you hooked your vacuum cleaner up to that kind of power, you'd melt its engine instantly—and possibly yourself as well. You previously read that transformers reduce the voltage before it enters your house. A few hundred thousand volts might sound like fun to your kids, but you should be grateful that you end up with a lot less voltage, thanks to transformers.

Know Your Volts: 120/240

Your utility company supplies your home with a split-phase 240-volt feed. This means that the two hot wires coming into your service panel are each carrying 120 volts. This provides power for major appliances such as electric stoves and clothes dryers that require both 240 volts and 120 volts. Just about everything else in your house only requires 120 volts.

Because of voltage drops in your house wiring, these numbers might drop as low as 220V and 110V. Some appliances and other electrical items even indicate on their instructions that they are rated at 220V or 110V. This is the manufacturers' way of telling you that their products will still work correctly when the voltage drops. An electrician or builder might refer to a 220/110 system, but it's really 240/120. This is an important concept. Reread this paragraph if you find it confusing.

Amps for Short

Volts measure the force of an electrical current, which is the movement of zillions of electrons. *Amperes* or *amps* measure the number of electrons in a current moving past a specific point on a wire or another conductor in a one-second period of time. If you want an exact number, one ampere equals one coulomb of electrical charge moving across a conductor in one second (or 6,250,000,000,000,000,000 electrons per second). *Coulombs* were named for that fun eighteenth-century French physicist, Charles A. de Coulomb. If electricity were water, volts would be the speed of the water, and amps would be the amount of water flowing through your hose.

A new three- or four-bedroom house often will have a 200-amp service; apartments or small condominiums might only need 100-amp services. Individual fuses and circuit breakers are measured in amperes. That is, they only allow a certain amount of current to pass through before they shut down the current.

Watt's That?

Watts are one of the measurements you will refer to in your electrical work. Most of us know the term from buying light bulbs (I know, the term should be lamps, but this isn't an easy one to get away from) as in "Why do we have a shelf full of 60-watt bulbs when I need 100?" A *watt* is a unit of electrical power. It tells you how much electricity you're consuming and being billed for by your utility company. This is why you have a meter outside your house recording your usage.

A single watt is a minuscule way to measure power usage, so the following larger units of measure are used instead:

➤ Kilowatt or kw (1 kw = 1,000 watts)

➤ Megawatt (1 megawatt = 1,000 kw or 1 million watts)

You've really got some usage problems if your electric bill indicates that you're in the megawatt range. (Perhaps you have a refrigerated warehouse on your roof.)

Wattage Around the House

How do watts figure into your electrical calculations? It's simple: They tell you how much stuff you can pile onto one circuit without overloading it. You don't want to put too much demand on a circuit with too many watt-hungry loads. Some simple math will keep you on the right track.

Watts equal voltage times amps. Let's say you want to install a new 15-amp circuit so you can add some outlets or lights to your living room and dining room. Because you won't be running any major appliances (assuming you don't do your laundry in the living room), this circuit will be running on 120 volts rather than 240. Therefore ...

$$120V \times 15A = 1,800 \text{ watts}$$

Terrific, you say. I can put in 18 100-watt lights. Actually, you can't, because you generally figure on running only 80 percent of the maximum load

Ask an Electrician

The **meter** associated with your electrical system is more correctly known as a **kilowatt-hour meter.** Your electric bill is calculated in kilowatt-hours (KWH). Each kilowatt-hour equals 1,000 watts of energy used for one hour. A 40-watt light bulb would have to burn 25 hours to equal 1 KWH.

(see Chapter 5)—1,440 watts in this case—but that's still 14 lights with watts to spare. What if you want to plug in your new window-shattering, guaranteed-to-have-the-neighbors-call-the-police music system that needs 1,900 watts all by itself? Time to re-calculate. It's going to need its own private 20-amp circuit before you can crank up Eric Clapton's original version of "Layla." By calculating your electrical needs first, you can accurately wire your house once without needing to make adjustments later.

Electrical Elaboration

Large appliances are on 240V lines because their power demands are so much greater than small items such as lights or clock radios. An electric range, for example, can demand more than 12,000 watts, and an electric dryer can draw close to 5,000 watts. Other major appliances such as refrigerators and washing machines only require a 120V line. They usually are assigned a separate circuit.

This Joule Isn't a Gemstone

Another oddball measurement you might run into if you buy, for example, a surge suppressor for your computer equipment is a *joule*. Named after James Joule, all-around smart guy and the son of a nineteenth-century English brewer, a joule in the field of electricity is the amount of energy equal to 1 watt acting for one second. When it comes to surge suppressors, the more joules of protection, the better.

Resistance Isn't Futile

Everywhere we look in life, we find some form of resistance. For airplanes, it shows up in the form of wind (and maybe an occasional bird or two). Water keeps kayakers afloat, but it also slows them down some. Even the indomitable James Bond in his Aston Martin DB5 had to contend with resistance when his tires hit the road.

It would be great if all the electrons in a current could go gliding across a copper wire (or another conductor) free and clear, but pesky resistance prevents them from doing so. Resistance in a conductor opposes the flow of an electric current. This results in some of the electrical energy changing to heat, which you want to minimize. Hot wires can be dangerous wires. On the other hand, resistance is built into the system to control the strength of the current running through it. You also don't want your blender getting hit with 50 amps of electricity when you're mixing a fruit shake.

Electrically speaking, resistance is measured in *ohms*. Ohm's Law (Ohm was a German physicist with a great name) basically says that the smaller the wire or conductor, the greater the resistance to a current. If you crank up the amps, you get even more resistance, sometimes to the point of overheating and causing a fire. Loads that require more amps also require larger wire to handle the current flow. If you increase the size of the wire, the resistance goes down, and you get a weaker current with less voltage drop. This is one of the reasons you have several different sizes of wire in your house.

Think of it this way: Imagine that the fire department is putting out a fire in your house. You're happy that they're using a big hose (just as a No.6 or No.10 is a big wire for big jobs). But what if you're just watering your garden? Then you want to conserve water and avoid flooding the garden. You'll use a small hose (just as a No.12 or No.14 wire is good for small items such as light fixtures).

All Wire Isn't Created Equal

Other than a brief foray in the 1970s when aluminum wiring was popular, copper is king when it comes to house wiring. Copper rates high on the conductivity scale. That is, it's an efficient pathway for an electrical current.

Ask an Electrician

The diameter of a wire is measured in **mils** or, more accurately, **circular mils** (**CM**). A mil equals $1/_{1,000}$ of an inch. Wire is assigned a **gauge value** by the American Wire Gauge (AWG) rating. The lower the number, the thicker the wire. No.8 gauge wire, for example, is thicker than No.14 gauge wire.

Conductivity Scale

Material	Percentage
Silver	100%
Copper	98%
Aluminum	61%
Iron	16%
Nickel	7%

In addition to a wire's conductivity, its size and the type of insulation around the wire affect its *ampacity,* or the amount of current (in amps) it can carry before it exceeds its temperature rating. The greater its size, as measured in *mils*, the more current it can conduct.

Every wire size has a maximum current that it can conduct. The following table shows the most common residential wire sizes and their ratings.

Wire Gauge Rating

Gauge Value	Ampacity
14	15 amps
12	20 amps
10	30 amps
8	40 amps
6	55 amps

Appliance and lamp cords use No.16 or No.18 wire, which is quite thin. This might lead you to ask, "Well, if thinner wire has a lot of resistance and can heat up easily, but thicker wire can hold more juice without overheating, why don't we use thicker wire throughout our homes? Wouldn't that be safer?" This is a reasonable question, and it has two answers: flexibility and cost.

Modern NMB cable, black or hot conductor; white or neutral conductor; grounding conductor; plastic insulation.

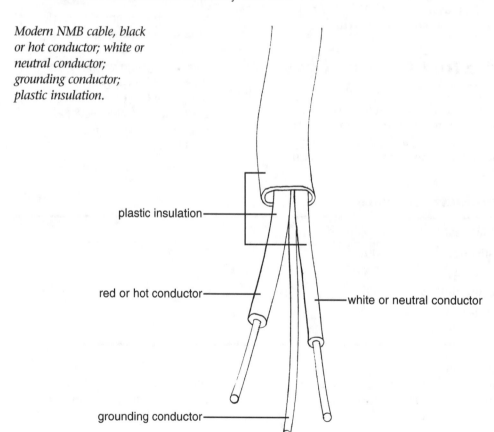

plastic insulation

red or hot conductor

white or neutral conductor

grounding conductor

If you try to bend and fit No.8 wire so you can connect it to a light fixture, you'll come to appreciate the flexibility of smaller wire. Like just about anything else in life, the larger the size, the greater the cost. There's a reason home-improvement stores periodically have loss-leader sales on No.12 wire but not any of the thicker stuff.

Other factors that affect your choice of wire will be discussed in later chapters.

No Substituting

The society of wire and conductors is a very closed one. No amount of politically correct persuasion will convince one gauge of wire to mingle with another. You should not mix No.12 wire with No.14 on the same circuit, for example. Wire must match up with its circuit breakers or fuses; No.14 wire doesn't go with a 20-amp breaker, so don't confuse either party by mixing them together. You *can* install larger wire on a smaller circuit breaker, but you cannot install a smaller wire on a larger circuit breaker.

The Least You Need to Know

➤ Electricity is a flow of electrons that, when in the form of an alternating current, can be used to power our homes.

➤ All modern electrical systems are grounded for safety.

➤ Volts, amps, and watts are common measurements of electricity.

➤ There are several fixed wire sizes. Each size of wire is designed to carry a specific amount of amps. Don't mix them up.

Chapter 3

History Lessons

In This Chapter

➤ The people who brought us electricity

➤ Safety—always an issue

➤ Evolving standards

➤ Knob-and-tube wiring

➤ Demanding more power

Many people think of the era of electricity as beginning with Thomas Edison and his electric lamp (or light bulb). His work was crucial in popularizing electricity and in making it practical for modern life, but a long list of scientists preceded Edison in this field. It took centuries of work just to discover what electricity is. I already mentioned the Greeks and their party tricks—creating static electricity by rubbing a piece of amber with wool or fur. They became pretty busy creating democracy as well as feta cheese, so they didn't get any further with electricity.

A couple *thousand* years later, around the year 1600, English scientist William Gilbert got the ball rolling again when he coined the term "electric" while describing the theory of magnetism. He was followed by a host of physicists, most of whom had laws, theories, or measurements named after them. These scientists laid the groundwork for industrialists like Edison and Westinghouse, who were able to exploit electricity and get it out of the laboratory.

Once the light bulbs started glowing, electricity became the computer industry of its day, with constant innovations, the building of an infrastructure, and a steady array of new uses. The dreams of merchandisers were realized in the years to follow, as they convinced the world to buy new gadgets and products that everyone previously had lived without, apparently in blissful ignorance. Electric lamps were followed by early versions of curling irons, electric cars, and waffle irons. Today, even Edison would be amazed at the electric world he helped create.

An International Effort

The development and nurturing of electrical power resulted from the work of scientists and accidental discoveries on both sides of the Atlantic Ocean. This essentially was a European and American deal, and it included contributions from England, Scotland, France, Yugoslavia, Germany, and Italy. The earliest attempts to create or reproduce electrical currents were through the use of crude batteries. The Energizer Bunny wouldn't have completed one drumbeat powered by these early batteries.

For the most part, these early physicists (they almost all were physicists) studied electrical phenomena, quantifying their observations so each one could conclude, "A-ha! It really hurts when you stand in a metal bucket of water and touch the bare ends of hot wires together!" Each contributor added to a gradually developing body of knowledge about electricity.

Positively Shocking

If you're ever playing a trivia game and are asked who invented the first electric light bulb (lamp), do not answer Thomas Edison! Sir Humphry Davy (1778–1829) conducted the earliest experiments in electric lighting, and the first incandescent bulbs were patented in 1840. Thomas Swan, another British inventor, was working on a carbon-filament lamp at the same time as Edison.

The Pioneers

A number of key players were poking and probing into electricity, most of them during the eighteenth and nineteenth centuries. Considering the unsophisticated equipment these scientists used, their accomplishments are that much more remarkable. It's not like they could refer to a textbook—they were writing the textbooks! You'll recognize some of these scientists as the namesakes of some electrical terms we use today.

Ben Franklin Flies a Kite

Ben Franklin, the colonial printer known for pithy quotes who is now pictured on $100 bills, is famous for having flown a kite during a lightning storm—a practice not advocated by this author or your local hospital. Franklin was testing his idea that lightning was a form of electrical current. A metal key attached to the kite attracted the lightning as its electrical charge traveled down the kite's cord and into Franklin's wrist. As a result of his

1752 kite-flying and his follow-up observations, Franklin developed the terms "conductor," "charge," "electrician," and not surprisingly, "electric shock."

Galvani's Frog Legs

Luigi Galvani had the perfect name for an East Coast Italian restaurant, but his only known association with gourmet food was his famous experiment with frog legs in 1786. The professor of medicine in Bologna accidentally produced an electric charge against the legs of a dead frog. The charge was the result of the wet frog lying on a metal plate while being probed with a knife made from a different metal. Galvani was convinced that the twitching legs were the result of electricity already existing in the frog's tissues and muscles. He was none too pleased when his friend Alessandro Volta disagreed and proved him wrong by showing that moisture caught between two different metals can create a small current, frog or no frog.

As a result of his disputatious observations, Volta went on to invent the first electric battery (called the voltaic pile) and, more important, to show that electricity could flow in a current along a wire instead of only in a single spark or shock.

In addition to being named a count in 1801 by Napoleon, Volta had the term "volt" named after him. As for Dr. Frog Legs, he walked away with the consolation prize of having the term "galvanism" (to have an electric current) named after him.

Watt

James Watt was an engineer at the University of Glasgow. He was a steam-engine guy who invented the steam-condensing engine and subsequent improvements in the 1700s. Watt was probably very motivated to work with steam: Scotland is cold in the winter, even with today's central heating. It must have felt like Antarctica back in the eighteenth century.

Bright Idea

You can impress your friends and family and probably drive them to vegetarianism at the same time by reproducing Galvani's discovery in your own kitchen. Instead of using a dead frog, try it with your Thanksgiving turkey right before you stuff it. You might have to cheat and wire it up to a hidden battery, but watch the fun when those legs start moving.

Edison coupled his own generator with Watt's steam engine to produce the first large-scale electricity generation. As you can guess, the term "watt," a unit of power, was named after Watt.

The Amp Man

André Marie Ampère, the first notable French electrophile, researched electricity and magnetism, essentially developing the field of electrodynamics. Not much for quotable sound bites, Ampère's most important publication, *Memoir on the Mathematical Theory of Electrodynamic Phenomena, Uniquely Deduced from Experience* (1827), is a book only a physicist could love.

A unit of electric current is called an "ampere" in his honor, but Americans, blatant and unapologetic in messing with the French language, call it an "amp" instead.

Oompa-Ohm

In 1827, Georg Simon Ohm, a German physicist and mathematician in Cologne, published *The Galvanic Circuit Investigated Mathematically,* a tome never destined to make the *New York Times* Bestseller list in any category. Lacking acceptance in his native Germany, Ohm eventually was awarded the Copley Medal in 1841 by The Royal Society of Great Britain. Ohm discovered one of the most fundamental laws of electricity: the relationship among resistance, current, and voltage. The resulting law, $V = IR$ (in which V is voltage, I is the current, and R is resistance), gave him a place in the electricity hall of fame. A unit of resistance, the ohm, is named after him.

Electrical Elaboration

Being the first kid on your block with a new law of physics didn't guarantee you stock options or, in the case of Georg Ohm, a pile of deutsche marks. As a university professor, Ohm was severely ridiculed when he tried to mathematically explain his theories of electrical resistance. He left his teaching post and lived for years in poverty before his theories were recognized and accepted. He lived the last years of his life as a full professor at the University of Munich.

Coulomb Was Très Cool

Charles Augustin de Coulomb was an all-around brilliant eighteenth-century French scientist who made major contributions in the areas of physics, civil engineering, and the natural sciences. The unit of electric charge—the coulomb—is named for him. Who could forget "Chuck" Coulomb's 1773 address to the Academy of Science in Paris when he discussed pioneering soil mechanics theory? Coulomb served as "Ingenieur du Roi" ("Engineer of the King") until the French Revolution came calling. He then took a powder and retired to the countryside for a while.

Coulomb is known in the electrical world for verifying the law of attraction or electrostatic force. Basically, he confirmed the notion that opposite charges (+ and –) attract each other and like charges repel. Unlike other observers of this behavior, Coulomb worked the numbers and came up with a nice, neat theory that no one outside the fields of electrical engineering and physics will ever use. Therefore, it's not worth mentioning in any detail in a how-to book, although it would have a real place in *The Complete Idiot's Guide to Physics*.

Other Electrical Fellows

The following European scientists also helped pave the way for the electrical comforts we enjoy today:

➤ Michael Faraday

➤ Heinrich Hertz

➤ Joseph Priestley

Joseph Priestley, in addition to his electrical dabbling, also invented soda water (for which the Coca-Cola Company is eternally grateful).

Michael Faraday is credited with discovering how to generate an electric current on a usable scale. It was known that electricity would create a magnetic field, but Faraday looked at the reverse notion: Why not produce electricity with magnets? In 1831, he discovered that moving a magnet inside a coiled copper wire produces a small electric current. If you spin a large enough magnet really fast inside a larger coil of wire, you'll have yourself a usable electric generator. Faraday's work is the basis for the electrical generators used today.

Now that we've discussed scientists from the Old World the colonists left behind, let's leap over to the American side of the Atlantic, where our usual combination of good timing, enthusiasm, and an attitude of "Hey, this will work, what have we got to lose?" put electricity on the map.

Ask an Electrician

The voltage in an alternating current goes back and forth between zero and maximum as it changes direction along a conductor. It then repeats this cycle 60 times per second, or 60 **hertz** (**Hz**), in ordinary house current. This measurement was named for the German physicist Heinrich Hertz.

Edison, Mega-Inventor

Thomas Alva Edison was born in Milan, Ohio, on February 11, 1847. According to some stories, he had a whopping three months of formal education, yet he obtained a record 1,093 patents in the United States during his lifetime. Who knows how he

Positively Shocking

On a more gruesome note, Edison was so adamant about maintaining direct current as the standard for electric power that he used the newly introduced electric chair to point out the dangers of alternating current. His demonstrations showed that a relatively small amount of current (the same current Westinghouse wanted to run inside of homes) could cause death or severe injury, unlike Edison's direct current.

Ask an Electrician

Incandescent refers to heating something until it is red-hot or white-hot and is glowing with heat, thereby giving off light. In a lamp, a filament with high resistance to an electrical current is used. Eventually, after being repeatedly heated to high temperatures, the filament breaks and the lamp needs replacement.

would have done without *any* public schooling! It seems like Edison had his hand in everything: telegraph equipment, movie projectors, phonographs, storage batteries, and most important, electrical lighting.

If Edison were alive today, he'd be spending most of his time in courtrooms defending his far-flung empire from charges of being a monopoly. Compared to Edison, Bill Gates is a piker. The list of Edison's companies and partnerships worldwide goes on for pages and pages. He not only manufactured electric lamps (a.k.a. light bulbs) but also motors, dynamos, phonographs and phonograph records, and telephone equipment. Edison helped form the nascent General Electric Company, one of today's powerhouse corporations, when his Edison General Electric Company merged with the Thomson-Houston Company. Despite his many inventions and businesses, Edison was only financially comfortable. He was nowhere near as wealthy as some of his contemporaries such as Henry Ford.

Let There Be Light

The basics of the construction of the electric lamp (or light bulb, to nonelectricians) were pretty well known by the 1870s. People knew that if you ran electricity down certain substances, the resistance produced light rather than heat. The problem was finding the right filament. Early versions simply didn't last long enough to be useful. The lamp needed a long-lasting filament that would provide pleasing, easy-on-the-eye lighting to be practical.

Edison tested thousands of materials before trying a piece of #70 coarse sewing-machine thread in October 1879. He first baked the thread to carbonize it and extend its life to withstand the heat of an electric current. The rest, as they say, is history. Edison and his assistants scrambled to improve his lamp and to create all the myriad components necessary to get it into peoples' homes. It was Edison's invention of a system to deliver and implement electricity and lighting that set him apart from other inventors. His labs designed and manufactured switches, meters, generators, and just about everything else connected with electrification. This is akin to inventing a computer in a laboratory only to discover that, oops, now we need software, monitors, a mouse, printers, scanners, and every other peripheral advertised in the monthly catalogs we all receive from computer suppliers.

Our First Big Power Plant

Edison designed and built his first major power plant in the Big Apple in 1882—a reported 120-volt system in downtown Manhattan. This also was the world's first principal power station. Unfortunately, Edison built it based on direct current, an approach that would become dated by the next decade, as was proven by one of his employees. Nikola Tesla, a Yugoslavian immigrant who briefly worked at the Edison laboratory in New Jersey in 1884, was on to something with his ideas about alternating current.

Con Edison, which started out as the New York Gas Light Company in 1823, is the current-day result of more than 170 mergers and acquisitions. The nucleus of Con Edison was the Edison Electric Illuminating Company, formed in 1880.

Tesla Needed a Lawyer

Nikola Tesla was another guy who liked applying for patents. By the time of his death in 1943, he held more than 700 patents in the areas of induction motors, generators, fluorescent lights, and steam turbines. Tesla supposedly arrived in America in 1884 with 4¢ in his pocket. (Who knows, maybe it's one of those stories that claimed a little less money every time Tesla retold it.) America's tough when you've only got 4¢ to your name. In 1885, Tesla sold his patent rights to his system of alternating current to George Westinghouse, another inventor and industrialist who knew a good electrical system when he saw it.

Tesla established his own laboratory in New York City in 1887. Ever the prankster, he sometimes would use his own body as an electrical conductor to light lamps to show that alternating current was safe. It probably was a great way to impress prospective girlfriends as well. While Westinghouse raked in the big bucks from his newly acquired alternating-current system, Tesla eventually

Bright Idea

Prescient investors in promising industries can reap huge fortunes as those industries grow. Looking back at the nascent electrical and phone industries, it's easy to see why computer technology, software, and telecommunications are good investments today. The trick, as always, is to know which industries and industry players will succeed.

became the namesake for a unit of measurement for magnetic fields. A *tesla,* as every amateur physicist knows, is equal to one *weber* (a unit of magnetic flux named after German physicist Wilhelm E. Weber, not the barbecue manufacturer) per square meter.

Considered both a genius and an eccentric during his lifetime, Nikola Tesla laid much of the practical and theoretical groundwork for the communications and electrical systems we have today.

Early Safety Measures

Electrical systems were a brand-spanking-new technology in the late nineteenth century, and no small amount of trepidation was associated with them. Wouldn't there be fires? Electrical shocks? Government officials, especially fire departments, took electrification very seriously.

The New York Board of Fire Underwriters, meeting in October 1881, called for standards such as the following:

➤ "Wires to have 50 percent conductivity above the amount calculated as necessary for the number of lights to be supplied by the wire."

➤ "Wires to be thoroughly insulated and doubly coated with some approved material."

➤ "Where electricity is conducted into a building from sources other than the building in which it is used, a shut off must be placed at the point of entrance to each building and the supply turned off when the lights are not in use."

➤ "Application for permission to use electric lights must be accompanied with a statement of the number and kind of lamps to be used, the estimate of some known electrician of the quantity of electricity required, and a sample of the wire at least three feet in length to be used, with a certificate of said electrician of the carrying capacity of the wire."

Electrical Elaboration

The first committee meeting for the National Electrical Code (NEC) was held in 1896, and the first code was published the following year. The NEC has adapted to the times and changes in electrical equipment, and it is now more than 1,000 pages long. More than 300 volunteer members of the National Fire Protection Association (NFPA) work on the code and code changes for new editions of the code.

These guys were serious! It's no wonder: They weren't about to take chances with a new technology that was potentially dangerous, despite its useful prospects.

The Standards Change

Once early wiring requirements were established they were stringent, but the size of an individual home service and the subsequent loads were inadequate by today's standards. A 30-amp service with a wood fuse box was typical and sufficient for running lights. Prior to electrification, many homes had gas lighting. Some wiring actually was fished through the gas piping in the walls to the new electric lights that replaced the old gas fixtures.

The turn of the century was a heady time in the field of electricity. Innovators and scientists were improving all the necessary components and generators as well as coming up with new gadgets and conveniences that would run on electricity. Even the newly built New York City subway systems were beholden to this great new power source.

Fuses to Breakers

All new residential construction uses circuit breakers in its service panels, but fuses were first used to distribute electricity through individual circuits. These fuses are still present in older homes where electrical services have not been updated. Although the first circuit breakers date back to as early as 1904, it wasn't until the 1950s, with the introduction of the modern plug-in-type breaker, that they gained universal usage.

Fuses might be dated and less convenient than circuit breakers, but they are perfectly usable under most circumstances. When updating to a new service panel, of course, fuse systems are always replaced.

Just One Ceiling Light

Compared to today's lighting standards, our grandparents and great-grandparents were almost walking around in the dark. A single ceiling light per room was considered an improvement over gas lighting. Can you imagine having only one overhead light in a kitchen? We bathe ourselves in light today, and we love every minute of it. The lighting requirements today in some kitchens alone would have consumed half of a typical home's service requirements back in the 30-amp days.

Positively Shocking

You would assume with the advent of electric lighting that Americans would have beaten a path to electricity's door. Not so! Improved gas-fixture innovations allowed for the continued use of gas lighting for domestic use and streetlights until after World War II. This is the equivalent of using a manual typewriter in the computer age.

Knob-and-Tube Wiring

The earliest wiring system was called *knob-and-tube wiring,* and you'll still find this in houses built prior to the early 1950s. This was an inherently safe system in which the hot wire and the neutral wire ran separate from each other through walls, floors, and ceilings. Each wire was covered with cloth insulation and ran through ceramic tubes when passing *through* floor joist and wall studs or into electrical boxes at lights, switches, and receptacles. The wires were secured to ceramic knobs when they ran *along* a joist or stud.

Electricians were a little anxious about this new electricity stuff, so most original knob-and-tube work is very neatly done. Wires were twisted together, soldered with lead, and then taped to make secure connections. The main problem with knob-and-tube wiring is what happens in the intervening years when homeowners and amateur electricians hack into it (see Chapter 6, "When You Buy a House").

Wiring Evolves

Although knob-and-tube wiring prevailed for years, other types of wiring were developed in attempts to either speed up or simplify installations. If your house was built prior to the 1950s, you might find one of these not-so-fun types of wire in it. Unfortunately, every innovation doesn't stand the test of time.

The following two systems were later contemporaries of knob-and-tube wiring:

➤ Armor-clad cable

➤ Multiconductor cable

Armor-clad cable, often called BX, was a trademark of General Electric. It consisted of a narrow metal sheathing wrapped around the hot and neutral conductors. It was mainly installed in the 1920s and 1930s in more expensive housing.

Running the wire in a protective metal wrapping sounds like a good idea, right? It probably was until the metal started rusting and corroding with age. Then the hot wire could short out to the metal wrapping, and in some cases, the wrapping could become red-hot but never blow a fuse. This type of wiring should always be thoroughly inspected by an experienced electrician.

Multiconductor cable carried both the hot and neutral wires in a cloth insulation that was coated with either varnish or shellac. Each wire was separately wrapped in its own insulation as well. As it ages, the insulation becomes very brittle and is almost impossible to work with in some cases.

Current systems use nonmetallic sheathed cable commonly known as Romex, another trade name. This cable is wrapped in thermoplastic, which probably will last so long that future anthropologists will be carbon-dating it in the year 14,500 C.E. This is a very safe wiring system that comes with a grounding wire in addition to the hot and neutral wires, and it allows for relatively quick and efficient installation.

Electrical Elaboration

Some early wiring was installed behind wood molding specifically made to house the wires. A millwork company would cut grooves on the backside of the wood, one groove for the hot wire and one for the neutral. This system had some drawbacks, the most obvious one being wire damage from nails driven through the wood. By the 1930s, wood molding was illegal and no longer was installed.

Creeping Home Power Demands

Early electrical and lighting systems were comparable to our more recent computer, software, and communications industries:

➤ The first electrical systems offered relatively little power and ran house- and streetlights. The first desktop computers had almost no memory, slow speed, and small hard drives.

➤ Innovation came fast and furious among very competitive companies and individuals.

➤ Our demand for more bandwidth, cable, and phone availability is a repeat of our increasing demands for more electricity since the turn of the century.

➤ All of these industries have improved our standard of living, despite criticisms to the contrary from technophobes.

A typical 30-amp home service has increased to 200 amps in the past 100 years. We've gone from one light per room to multiple lights, multiple receptacles, every appliance imaginable, and entertainment systems all demanding their share of electricity. This has required building and rebuilding an entire infrastructure of dams, generators, long-distance power lines, transformers, and miles and miles of utility poles.

The entire undertaking has been enormous and is entirely taken for granted today. Historical hubris lets us believe that our time is the most innovative and influential to date, but we wouldn't have gotten very far without electrification. Try running your laptop on some of Volta's original batteries or even some of Edison's. You might get enough power to read "Starting Windows 98" on your screen before it shuts off, with your battery drained of any direct current.

The basics of electricity and its delivery systems are pretty well established. Equipment might improve and become more efficient, but until someone rewrites the laws of physics, electricity will continue to be delivered by wires or other conductors from a generating force. You'll still get billed once a month or so for its usage. Nobody said all those electrons would be free, but it remains quite the bargain based on all it provides for us.

The Least You Need to Know

➤ The history of electricity and electrification is full of pioneering scientists, each building on the knowledge of the others.

➤ Thomas Edison's biggest contribution was building usable wiring and lighting systems.

➤ Early electrical systems used knob-and-tube wiring and fuses—both still found in older homes.

➤ A century of innovation has brought us safe, efficient power and power-delivery systems far beyond the dreams of industry pioneers.

Chapter 4

If Your Walls Could Talk

In This Chapter

➤ National Electrical Code requirements

➤ Inspecting your home for compliance

➤ Understanding wiring systems

➤ Understanding circuit breakers and fuses

➤ Installing plenty of power

Many horror movies feature haunted houses with fiends, evil spirits, or some other form of calamity lurking behind the walls or under the floors. Bookshelves conceal hidden passages, and trap doors spring beneath the secondary characters, just missing the hero or heroine and plunging them to an unknown fate. Your house isn't anywhere near as fateful—at least it shouldn't be—but it does have some tales to tell behind the drywall or plaster, the walls you stare at every day.

Houses are built according to the building codes at the time of their construction. Even an owner-built house has to meet all code requirements including electrical requirements. As codes change so do construction standards. Any remodeling work you do now has to meet the current code, even if the rest of your house does not.

If your house is old enough, it probably has been remodeled or altered on a number of occasions. It's likely that some of this work is, putting it kindly, code-challenged. You've already taken a superficial look at your wiring. Now let's examine it in more detail. A good starting point, so you'll know what all the fuss is about, is the National Electrical Code.

The National Electrical Code

The National Electrical Code (NEC) is the guiding light behind most electrical installations in the United States. As a rule, local regulations include the NEC along with any specific ordinances imposed by individual building departments. The fact that so many state and city agencies use the NEC affirms the excellence of this code. As a culture, we might have our disagreements from one coast to the other, but we do agree on the National Electrical Code. Interestingly, the NEC is not mandatory and therefore has no regulatory control. It's strictly a set of voluntary guidelines, which is a good thing: If Congress had gotten hold of it and tried to turn it into legislation, we'd still be using candles and kerosene lamps to light our homes.

The NEC is just over 100 years old, and it has changed with the times. The advent of electrical power brought with it the need for standards for equipment, installations, and usage. The initial concerns around electricity were mainly about property protection. Fire underwriters were especially concerned, because they were taking losses due to electrical fires.

New York, the late-nineteenth-century center of the universe, wrote the first requirements for electrical safeguards in 1881. More codes by different industries followed, and in 1892, the Underwriters National Electrical Association consolidated these various codes, although there was still no universal acceptance of one set of rules. Saying enough was enough, the National Electric Light Association called a conference in 1896 to come up with one standard set of rules. The conference included utility representatives, underwriters, inspectors, and just about every key player who was involved with this new electricity business. They diced, spliced, and blended the best parts of all the existing American and European codes. They then tweaked it during a review process, and in 1897, they came up with the National Electrical Code. There have been 48 editions of the code, and the next one is due out in 2002.

The National Fire Protection Association (NFPA) has administered the NEC since 1911. Harking back to the original NEC, the current code is a product of group consensus with an even greater variety of contributors including labor unions, testing laboratories, regulatory bodies, and consumer groups.

The NEC goes well beyond home electrical systems. Among other things, it covers the following:

➤ Fiber-optic cables

➤ Antennas

➤ Fire alarms

➤ Marinas

Bright Idea

You don't necessarily need to study the NEC itself to do your electrical work. Most of the rules that apply to home electrical work are commonly known and are discussed in books such as this one. If you have any questions or concerns, point them out in your plans when you apply for a permit and get a ruling before you do the work.

➤ Carnivals, circuses, and fairs

➤ Mobile homes

➤ Gasoline dispensers

Espresso stands and popcorn wagons might not specifically be named, but they're covered, too!

Electrical Elaboration

When I was a kid, my family vacationed at Catawba Island on Lake Erie. (This was Ohio's version of a waterfront resort.) Near our hotel, there was an arcade full of hokey games and amusements that were designed to separate us from our saved weekly allowances. We discovered that, if you put one hand on the metal plunger of the pinball machine and the other hand on the metal casing of the mechanized gypsy fortuneteller, you got a fingertip-to-fingertip shock. This was great fun for endurance contests and for tricking unsuspecting cousins, but it left something to be desired from a safety standpoint.

You and the Code

The NEC exists for your protection. I know we live in an era of business-guru authors and self-improvement advisors extolling everyone to "think outside the box," to develop new paradigms, and to seek out the "wow" in everything we do. That's fine, just don't try it with your electrical work. The code will help keep you and your family alive and your house from burning down. You can be as imaginative as you want when it comes to selecting fixtures or adding lights, but they have to be wired and installed properly.

Just in case you decide to seek your bliss and express yourself with electrical work that isn't up to code, your local inspector will be equally expressive, but in a way that won't make you very blissful.

Local Rules, Local Inspectors

Ultimately, the NEC isn't the final judge and jury of your electrical work. This role belongs to your local electrical inspector who enforces your local electrical code. How does your local code differ from the NEC? It depends on where you live. Some municipalities stick with and solely enforce the NEC; some have additional rules. Remember, your local code determines how you or your electrician will do your electrical work, so you must know the regulations.

The electrical inspector wears a variety of hats. An inspector …

➤ Interprets the NEC rules and regulations.

➤ Approves or rejects electrical work.

➤ Approves fixtures and materials.

Is an inspector always right? Let me put it this way: If you question a judgment or ruling, you had better be able to back it up by quoting chapter and verse from the code. No one is perfect, and an electrical code is an involved and complex document. Mistakes and misinterpretations are made on both sides, but the inspector has the final word. Take some tips from Mr. Etiquette:

➤ Do your best work, and do it neatly.

➤ Don't try to hide anything or take shortcuts.

➤ If your inspector believes you've erred in your work, listen politely and see what you have to do to resolve the problem.

Positively Shocking

Your insurance company can deny any damage claims to your house caused by uninspected electrical work. An inspection proves to your insurance company that the work was done according to the electrical code. If for any reason you do have an electrical fire, at least you can show the workmanship wasn't to blame.

It isn't the inspector's job to show you how to do your work. As a homeowner, you'll have to establish your credibility and competence to do the job, even more so than a trained electrician. An inspection is like the speed limit—you might not like it, but it's there to protect you.

The CEC

Canadians use the Canadian Electrical Code (CEC). You might wonder what this has to do with electrical work on this side of the border, but Alaskans (state motto: It Doesn't Get Any Colder Than This) have had problems in the past with Canadian-built outdoor work modules (portable buildings). Differences in the two electrical codes have resulted in some modules being unacceptable without costly upgrades to meet the NEC. You could probably argue that Alaska should be part of Canada, but it isn't, so the CEC isn't acceptable there.

Electrical Elaboration

A real-life example of a change in the National Electrical Code that has prevented human fatalities is the requirement to install ground-fault circuit interrupters (GFCIs). Statistics have shown that the installation of GFCIs in bathrooms, kitchens, the outdoors, and other potentially damp, hazardous locations has saved lives.

Underwriters Laboratory

Underwriters Laboratory is the organization that brings us the ubiquitous "UL" tags on just about everything we plug in or turn on. What does this group do? In the organization's own words, UL "is an independent, not-for-profit, product-safety-testing and -certification organization." Underwriters Laboratory was established in 1894 to test products for the emerging electrical industry. It has managed to attach more than 14 billion UL tags to products all over the world.

The UL tag is known 'round the world.

UL is *the* leader in product testing and certification. If a manufacturer wants its new gadget to gain ready consumer acceptance, it applies for the UL tag, because no retail outlet in its right mind would sell anything electrical without it. You can find the organization at www.ul.com on the Web.

The UL, along with the Consumer Product Safety Commission and the National Electrical Manufacturer's Association, helped establish the not-for-profit National Electrical Safety Foundation (NESF) in 1994. The Foundation's mission is to improve our awareness of electrical safety at home, work, and school. The NESF's Web site, www.nesf.org, points out the following cheery statistics:

➤ One person is electrocuted in his or her home every 25 hours, and more than 350 people die in over 40,000 residential electrical fires every year (Consumer Product Safety Commission data).

➤ One worker is electrocuted on the job every day (Occupational Safety and Health Administration data).

➤ Personal-property damage from fires exceeds $2 billion a year.

These are not-so-subtle hints that you can't take electrical safety for granted. The NESF offers tips, Web links, and a safety booklet, and it's all free for the asking.

Are You Up-to-Date?

Now that you've got a hint of the hell-and-brimstone that awaits you if you ignore electrical safety and code requirements, you can take a fresh look at your own system. As we've already mentioned, a new system can almost be ignored until you're ready to add to it. A system meeting the 1996 or 1999 NEC will be grounded, will have plenty of receptacles including GFCIs, and will have power properly distributed among a series of circuits throughout your house.

Positively Shocking

It's not an uncommon practice to do home remodeling without permits and inspections. Common acceptance doesn't mean it's advisable, especially when doing structural or mechanical work. I know of one seller who was later sued when his illegal remodeling came to light after the buyer took possession. Lawsuits aside, you don't want to be on the receiving end of unsafe work.

What if you have an older system or one that's been altered? Your home might have had an addition or two put on by past owners. How can you be sure these were done according to past building codes? Let's start with some basics as you review your electrical system.

There are some basic differences between older and newer electrical systems, such as …

➤ The presence or absence of a grounding conductor.

➤ Fuses versus circuit breakers.

➤ Nonpolarized outlets versus polarized outlets.

Two-Wire and Three-Wire Systems

A two-wire system means you don't have a grounding conductor. This is the bare copper or green-insulation-clad wire attached to a *grounding electrode* outside your house. Alternatively, the *grounding conductor* can be attached to a water pipe near your electrical panel.

Your first clue to determine whether your system has a grounding conductor comes from your receptacles. A two-pronged receptacle usually doesn't have a ground; a three-pronged outlet *should* have a ground. An older house might well have three-pronged outlets, but there is no guarantee that they were installed with grounds (another peril of uninspected remodeling work). To be certain, you must either …

➤ Test each outlet with a circuit tester, which will indicate the presence of a ground wire.

➤ Take the cover plate off each receptacle and look for the ground wire.

Of the two methods, using a circuit tester is easier (see Chapter 7, "Caution Signs and Safety Concerns"). If your system isn't grounded, it doesn't mean you're in grave danger and should refrain from turning on the lights ever again. It does mean, however, that it's a dated system that lacks a modern safety feature—a safety feature that people lived without until the 1960s. Unless there's been a deep conspiracy to cover up massive electrocutions of homeowners over the years, you can still live with an old two-wire system (but it's always a good idea to upgrade).

> **Ask an Electrician**
>
> A **grounding electrode** is a metal pipe or rod, often iron or steel-galvanized, that is driven at least eight feet into the ground outside your house. The material must be at least $5/8$ inch in diameter and must offer a resistance of 25 ohms or less. The **grounding conductor** carries any excess voltage to the ground to prevent injury or damage to electrical equipment.

A grounded receptacle connects any exposed metal sections of an appliance or lamp to the house grounding system. This means an errant current shouldn't pass through the metal shell of your washing machine and turn your wash day into something unexpected.

A three-pronged, grounded receptacle is the made-to-fit receiving end for a three-pronged plug. Remember, electrical systems are nice and logical. If you have a two-pronged outlet and a three-pronged plug, they don't go together, no matter how hard you push on the plug. Before you ask, cutting off the grounding pin from the plug is a bad idea. Buy an adapter, which is available at any hardware store, instead.

Feeling Polarized

Have you ever noticed how some receptacles (and all new ones) have two different-size slots? These are polarized receptacles, and they're the perfect fit for polarized plugs. In each case, one side (the slot of the outlet and one prong of the plug) is larger than the other. This is not done to intentionally annoy you, but if you have an old house, it can be a real inconvenience when plugging in new electrical devices.

A modern three-pronged, grounded, and polarized receptacle.

(Photo courtesy of Leviton Manufacturing Co., Inc.)

Bright Idea

A two-pronged plug adapter used to convert three-pronged plugs for use in older receptacles also can be used for adapting polarized plugs. The adapter enables them to be plugged in to nonpolarized outlets.

A polarized receptacle is constructed so a polarized plug can be inserted only one way. It also ensures that the hot leads and the neutral wires line up from outlet to plug. It might be tempting to file down the larger prong on a polarized plug, but trust me, this is a bad idea, too. These are all parts of a system, and they're made to work together and be kept intact. This is no place for quick fixes. Either adapt to the limitations of your older electrical system, or bring it up to date.

Circuit Breakers vs. Fuses

Circuit breakers and fuses perform the same task: They interrupt electrical power when the current demand is too high. If you plug a 2,200-watt personal surfing pool complete with Oahu-inspired monster waves into a 15-amp circuit, be prepared to wipe out along with the power. If the breaker doesn't trip or the fuse burn out, the wire connecting the receptacle to the panel could overheat and possibly start a fire. You probably wouldn't realize anything was burning because the wiring is concealed inside your walls.

Circuit breakers have been the mainstay for residential electrical installations for more than 40 years. They're convenient, dependable, and reusable. What's not to like about them?

This One Blows

Although there's a variety of fuses, most old residential systems use the glass, screw-in type called *plug fuses*. These fuses feature a narrow metal strip, visible through the glass, that quickly melts when too much current is starting to move through the circuit. Fuses are rated by amperage and cannot be reused after they burn out or "blow."

A fuse system generally is safe if it's used according to its design. This means …

➤ Always using the correct size fuse.

➤ Being sure the fuse is screwed in tightly.

➤ Not listening to your old uncle Bob, who says you can replace a fuse with a Lincoln-head penny and call it good.

The glass of a blown-out fuse might look smoky, or more noticeably, the strip will be melted or separated. Before replacing it, you should turn off any electrical load that might have caused the overload. Some electricians recommend that you also turn off the main switch to the fuse box before removing and replacing the old fuse. Whatever you do, *be sure to keep one hand behind your back or in your pocket!* You don't want both hands near the box, because if both accidentally touch it, you can become part of a circuit and electrocute yourself.

Bright Idea

Always keep spare fuses on hand, at least for the common 15- and 20-amp sizes. Keep the fuses stored with a flashlight so you don't trip around in the dark while replacing a burned-out fuse.

This One Trips

Breakers, as previously explained, "trip" when a metallic strip heats up, bends, and forces spring-loaded contacts apart. After tripping, they can be reset by pushing the breaker's switch all the way "off" and then back over to the "on" position. When resetting a circuit breaker, be sure to …

➤ First find out what caused the breaker to trip and shut it off.

➤ Use only one hand and keep the other one away from the service panel.

A circuit map will identify the circuits controlled by each breaker or fuse (see Chapter 5, "More Wall Talk").

Plug fuse; circuit breaker.

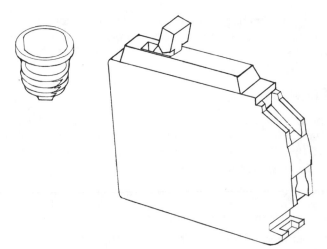

Circuit breakers, like fuses, are rated by amperage. Don't even consider installing a higher-rated breaker in place of a lower one. If you have repeated tripping, you're overloading the circuit and need to change your usage habits or upgrade the circuit.

Bright Idea

If you repeatedly trip a circuit breaker or burn out a fuse, you need to decrease your power demands. Examine your usage. You might be using a high-wattage appliance such as a hair dryer or a plug-in heater in addition to the circuit's normal load. Do a load calculation and determine what you'll have to change.

The Main Shutoff

Fuse boxes and service panels both have some kind of main shutoff mechanism, and you should know how to use it. The NEC says that you cannot have more than six disconnects—no more than six fuses, breakers, or levers of some kind—to turn off the service. Fuse boxes usually have either a single pull-down handle or two 50-amp pullout fuse holders to disconnect the service.

In a service panel, you will have either a main breaker or a series of breakers in the top half of some older panels that, when shut off, will disconnect the service. Remember, even with the breakers shutoff or the fuses removed, without the meter removed from the meter socket, the panel is still hot where the large wires come in from outside your house.

In case of an emergency in which you must get the power off quickly and cannot identify the specific fuse or breaker, the main shutoff will do the job. Keep in mind that your entire house can go dark. Be sure to have a flashlight close to your fuse box or service panel.

The More Power the Better

Harry Truman once made a comment about fully equipped armed forces, saying something like, "If we've got 'em, then we won't have to use 'em." He figured the more power the country had, the more of a deterrent it would be to future aggressors (if only he'd been right). He was on to something about power, though: It's better to have plenty of it than to not have enough. The same is true with your electrical system.

There is nothing wasteful or ecologically sinful about having a minimum 200-amp service in your house. No one says you have to use all that electricity all the time or even ever. Heed the Boy Scout motto: Be prepared!

Look to Future Needs

None of us knows what the future will bring, except maybe dial-up psychics: They know your call will bring them $2.95 a minute. Everything else is a roll of the dice. You might add a second story to your house, set up a woodworking shop in the basement, or double the size of your kitchen. You even could add a casino extension off your garage if your state legislature decides that home-based gambling operations are more in keeping with its ideas about family values.

Any and all of these possibilities require electric power. If you're replacing your current system, think more rather than less because, in this case, less is definitely not more.

Cost Comparisons

Trying to provide local price comparisons in a nationally sold book is like judging a national chili contest: What works in Massachusetts isn't going to cut it in Texas. In other words, there are a lot of variables, including …

➤ Local labor prices.

➤ The brand of service being installed.

➤ The difficulty of an individual installation.

➤ Permit and utility fees.

That said, we can make some generalizations based on the size of the service alone. The most common sizes of services are 100-, 125-, and 200-amp *single-phase,* three-wire systems. The approximate difference in cost—and you should always check with your local supplier or electrician—between installing a 100-amp service and a 125-amp service is around $100 plus permit and utility fees. The difference between a 125-amp service and a 200-amp service is around $150. The difference will increase depending on the length of conduit and wire required for the installation.

Based on a $150,000 house, the extra $150 equals one tenth of 1 percent of your home's value, or an espresso a week for one year. If it's a budget consideration, skip the coffee and go with the bigger service.

Ask an Electrician

A **single-phase** system or line provided by your utility can carry electrical loads suitable to meet the needs of residential customers and smaller commercial clients. A **phase** describes the voltage or current relationship of two alternating waveforms as the electricity travels down a conductor or, in this case, a wire.

Positively Shocking

If you have an older service panel, it should be checked by a licensed electrician to make sure it's safe. Some panels have shown regular safety problems and are not considered reliable by many electricians.

Panels and More Panels

Every electrician has favorite electrical components to recommend, and this includes service panels. Electricians also have panels that they avoid and existing, older panels that they recommend tearing out yesterday if not sooner. This is touchy ground, and anything I or the electricians I have worked with recommend will be subject to criticism in some quarters. With that caveat, I can recommend the following guidelines for choosing a panel:

➤ You get what you pay for, so skip the low-end panels.

➤ Ask as many electricians, electrical suppliers, and builders of high-end homes as you can what panel they recommend; one or two names should keep coming up.

➤ Buy the best panels made by your manufacturer of choice rather than lower-end units made to compete with similar low-end units from chain discount stores.

Consider what a panel does: It acts as the major line of defense in the event of an electrical problem. You want the best panel possible rather than the most basic. They all will be UL-approved, but it's better to go with a panel that will meet higher standards than the bare minimum. You wouldn't buy the cheapest brakes for the family minivan, and this is the same attitude you should take with your electrical panel.

What does my electrician recommend? He uses the Square D QO panel. Others recommend Cutler-Hammer or Siemens. All of these are reliable brand names that will serve you well.

The Least You Need to Know

➤ The National Electrical Code (NEC) is a set of guidelines for electrical installations.

➤ The final judge of your electrical work is your local electrical inspector.

➤ The safety features in a newer, three-wire electrical system include circuit breakers, a grounding conductor, and three-pronged, polarized receptacles.

➤ For a small difference in cost, you can install a larger electrical service that will more easily take care of future needs and remodeling.

Chapter 5

More Wall Talk

In This Chapter

➤ Understanding how circuits work

➤ Ground-fault circuit interrupters

➤ One house, different conductors

➤ Living with old wiring

➤ Wiring hazards

➤ An old-house checklist

Now that you know about fuses and circuit breakers, it's time to familiarize yourself with the circuits they control. A circuit is a continuous loop of electrical current. It travels from a power source (your utility through your service panel or fuse box) along a "hot" conductor (the black or red wire) as a load demands it. The electricity then travels back to the panel along the white or neutral wire. When you activate a load by flipping a light switch or turning on a food processor, you allow the circuit to pass through and light up the room or grind up basil leaves to make pesto.

Circuits aren't just random pathways for a bunch of juvenile-delinquent electrons out to cause trouble. They are determined by voltage or electromotive force (EMF). Amperage is the rate at which these electrons flow, and it's held in check by the size of your circuit breaker or fuse and the gauge of the wire. The size of the wire used in an individual circuit is influenced by the loads it must supply and even the distance the electricity must travel. Some circuits are dedicated. That is, they only control a single load such as a clothes dryer.

Your circuits run along conductors or wire, which usually is made from copper. Over the years, different wiring systems have been used. An older home that's been remodeled might have more than one type of wire or cable, and these can be a problem if they are not properly connected or are overloaded. This chapter will give you a better idea of what shape your electrical system is in and what you need to consider before altering it.

Branching Out to Break Up the Load

You could connect all of your 120-volt electrical loads—lights, bedroom receptacles, the refrigerator, and so on—to one big circuit breaker, and they would still function. Replacing a broken switch or installing a new light fixture then would mean turning off all the power to the house rather than just the power to one particular room. If your housecleaning service trips this giant breaker with a new 55-horsepower industrial vacuum cleaner while you are out of town and is afraid to reset it, you can kiss that frozen Copper River salmon in the freezer good-bye.

Electrical current is broken down into individual circuits—called *branch circuits*—for safety and convenience. You don't want the entire house to go dark because a GFCI in the kitchen tripped due to a faulty appliance. Each circuit is designed to carry a certain amperage and to provide enough current to meet the wattage demands of receptacles, lights, and appliances. The following figure shows a standard residential distribution of circuits.

A circuit is laid out logically, or at least it should be. This means that a 15-amp lighting circuit will control lights in, say, three continuous rooms rather than in three rooms at opposite corners and on different floors of the house. Several forces work together to help a circuit do its job safely.

Bright Idea

It's usually better to have more circuits than to have fewer, even if some of them seem underused. This way, you have extra capacity for future expansion without having to run entirely new circuits.

Ask an Electrician

A **branch circuit** is separate and independent from any other circuit. It has its own circuit breaker or fuse, wiring, and loads. Your service panel has a number of branch circuits for both 120- and 240-volt loads.

Amps, Watts, and Wire Gauge Working Together

You remember that amps, or amperes, are a measure of an electrical current's strength or flow. A watt measures the electrical power itself. That is, it measures the amount of electricity consumed by an appliance or another fixture as it converts the electricity into something useful to us. A circuit is sized to allow a certain amount of electricity to run to a given number of loads. The loads are measured in watts, which is why you can only have a certain number of receptacles and lights on a circuit. Too many running at once demands more juice than the circuit can safely provide before a protective device in the form of a circuit breaker or fuse steps in like a responsible bartender and cuts you off.

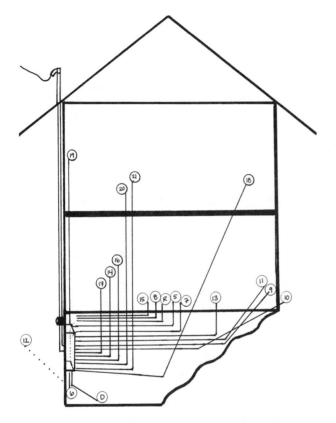

Typical branch circuits.

The amount of current carried to the various loads also is determined by the size of the wire running between the service panel and the loads. If your wire is too small for the amount of current the load is demanding, it will have a high resistance and will overheat. This is okay for the heating element in your toaster but not for your house wiring. The Wire Gauge Rating table in Chapter 2, "What Is Electricity Anyway?" shows the ampacity rating for various gauges of wire. A 20-amp circuit, which usually runs small appliances, requires No.12 wire. Many electricians recommend No.12 wire as the minimum size wire for residential use, even though the code accepts No.14 wire as the minimum-size conductor for branch circuit wiring.

Circuits and Runs

Circuits can be divided by type:

➤ General-purpose or lighting circuits

➤ Dedicated circuits

➤ Small-appliance circuits

Lighting circuits include most receptacles in living areas other than the kitchen, bathrooms, and workrooms. This is appropriate for most receptacles because we generally use them for small loads such as floor and table lamps or clock radios. Some receptacles will only be used for night-lights; others might rarely or never be used. Even light fixtures have varying loads depending on the wattage of their lamps. They can vary from 25 watts to 150 watts.

What happens when you plug in something larger such as a room air conditioner? What if you have a water heater or an electric range that also requires large amounts of current? These loads call for dedicated circuits, which are so-named because they only supply power to one specific load.

Bright Idea

Loads that demand large, intermittent amounts of power, such as a table saw in a workshop, also should be run on dedicated circuits. Power tools can easily trip a circuit breaker when they're first switched on if the circuit already is in full use from other loads.

Dedicated circuits include those for ...

➤ Major appliances.

➤ Refrigerators.

➤ Computers.

It's easy to understand why a major appliance needs a dedicated circuit, but what about refrigerators and computers? Even a large refrigerator-freezer combination is rated at about 500 to 700 watts, and a computer is far less. (My notebook PC is a minuscule 36 watts.) These fall into a different category of dedicated circuits that aren't based on a demand for electrical current but on their specific activity. It isn't critical for your refrigerator to be on its own circuit from a power-demand standpoint, but if another load somewhere else on the circuit's run trips the entire circuit, your refrigerator will shut down, and you will be looking at a lot of spoiled food. Some jurisdictions' codes require that the refrigerator be on a dedicated circuit.

Computers don't store food, but they do store your data. Most people readily agree that we should back up and save our documents and spreadsheets while we're working on them, but then we cheerfully continue working without doing either.

If a breaker trips because that lamp you keep meaning to rewire finally shorts, it's good-bye to the last chapter of the Great American Novel you've been working on for the last five years. Whenever practical, it's a good idea to install a dedicated 20-amp circuit for your computer and its peripherals.

Know Your Circuits with a Circuit Map

Electricians follow the minimalist school of writing: They write as little as possible when listing the circuits on the form inside the service panel door. An electrician will write "Lighting circuits," for example, across the space designating one or two specific breakers. That's all well and good, but a more useful description would say, "Ceiling lights in master bedroom and north bedroom, in second-floor hallway, and at top of stairs." These written descriptions need more room than most factory-supplied lists provide, unless the writing is very small.

Electrical Elaboration

Developing a circuit map is more than simply an exercise in knowing which fuse or circuit breaker controls which loads. It also can make you aware of overloaded circuits that have had too many loads added to them by misinformed homeowners. If a circuit appears to be overloaded, do a load calculation and determine if a load should be removed. Most likely, the problem will be an add-on done after the system was installed and probably done without an inspection or permit.

In a new house, the electrician's list usually is adequate because the wiring is so straightforward. In an old house, however, the list needs to be more specific, especially if past homeowners have added their own electrical marks when they lived in your home. Every owner has different needs, and they manifest themselves with receptacles, lights, and switches in places that will make no apparent sense to you but are perfectly logical for someone else. These include lights in crawl spaces, switches on attic rafters, and receptacles in closets. You don't need to know their history, but you should know which fuse or circuit breaker controls them, especially if they are tied in to the middle of a circuit and can potentially cause problems.

Drawing your own circuit map can be done alone, but it is best done with some helping hands. To come up with your map, you'll need the following:

➤ Paper and pen for recording

➤ Lights and radios to plug into receptacles

➤ Extra people spread around your house

What Are Friends For?

You'll be testing your circuits one at a time. It's just a matter of flipping a breaker or removing a fuse, observing what goes out, and writing it down. This means every light will have to be turned on, and every receptacle will need something plugged into it. This is where the lights and radios come in. It's important that you know where every receptacle and odd light is located and that you account for all of them.

Positively Shocking

In older homes, homeowners and do-it-yourselfers sometimes play fast and loose with good wiring practices and connect a new wire to the nearest convenient old wire. This can result in some odd circuits that might take longer to figure out when you make up your circuit map. Don't leave any receptacle or light unaccounted for just because it doesn't turn off when other loads in the room do.

As previously mentioned, this job is easier to do if you can fill up your house with some friends. As you turn the power off one circuit at a time, they can yell out what went off, and you can write it down in your notebook. This can be a fun project in a big house with people running and shouting as lights go off all over the place. Carefully write down everything as specifically as possible and redo the list later on your computer. Print a list to attach to your service panel or fuse box, and maintain the file on a disk to record new circuits as they are added or old ones as they are changed.

GFCIs—Ground-Fault Circuit Interrupters

The code requirement to install GFCIs has been a real lifesaver for homeowners. One government statistic suggests that a GFCI installed in every home in America could prevent more than two thirds of all residential electrocutions. A GFCI measures the current flowing into the outlet through the black or hot wire and the current outflow through the neutral or white wire (see the following figure). If the GFCI detects any difference greater than 7 milliamps, it shuts off the current. Why? Because any difference in the current is an indication that the current is somehow shorting or "leaking"—maybe through you! It might be a short in an appliance such as a hair dryer or an electric mixer. These are dangerous situations, and a GFCI will shut down far faster (in as little as $\frac{1}{40}$ of a second) than a standard circuit breaker or fuse.

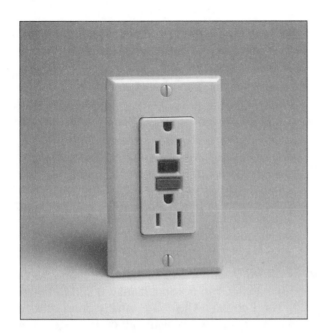

A GFCI.

(Photo courtesy of Leviton Manufacturing Co., Inc.)

An older home may or may not have GFCIs, but it can be retrofitted with them (see Chapters 19, "Kitchen Power," or 20, "Bathroom Wiring").

A GFCI won't always prevent an initial shock, but it does prevent a lethal one. The current NEC calls for GFCIs to be installed in a number of locations, including …

➤ Bathrooms.

➤ Kitchen counters.

➤ Outdoors.

➤ Garage walls.

➤ Unfinished basements and crawl spaces.

Why have these areas been singled out? They all have something in common: water or water pipes, both of which are good at seducing a current away from its righteous path back to your panel or fuse box to take a trip through your body instead. There's a reason why various cads in the movies get knocked off while in the bathtub when an irked female character throws a plugged-in curling iron into the water. If you use a defective hair dryer when touching the water faucet in your own bathroom, you'll be glad you have a GFCI installed.

A modern electrical system will have GFCIs in the form of either outlets or breakers, although the latter are more expensive. The presence of a GFCI in an old system is a demonstration that a past owner was concerned enough to attempt at least a partial modernization of the system. GFCIs are covered more completely in Chapter 11, "Switches and Receptacles."

Bright Idea

Test all your GFCIs monthly. It's simple and quick, and it ensures that the outlets are working properly. Plug a small night-light into the outlet and press the "Test" button. The light should turn off, and the "Reset" button should pop out. This trips or disengages the outlet. Press the "Reset" button to complete the test. If the "Reset" button does not pop out, you might have a defective GFCI, and it should be inspected or replaced. Note that a GFCI installed on an ungrounded circuit will not trip when the "Test" button is pressed. The "Test" button will only work if the GFCI receptacle is grounded.

Wire Systems Old and New

A main difference between old electrical systems and new ones is the presence of a grounding conductor in contemporary wiring. Modern cable contains all three wires—hot, neutral, and ground—wrapped in protective thermoplastic insulation. Your house might have cable running through it rather than the old knob-and-tube or metal-wrapped BX cable systems, but this is no guarantee that it's got a grounding conductor. Cable installed in the 1950s only contained a hot wire and a neutral wire. The only way you'll know for sure with your own cable is to check your electrical panel (see Chapter 17, "Service with an Attitude").

You've already read that old wiring isn't necessarily bad wiring, with the (sometimes) exception of BX cable, which can become damaged and be conducive to short circuits. The abuse usually occurs later as it gets hacked into. Replacing an entire system is expensive. New services installed in older homes usually incorporate some of the existing wiring into the new panel (unless it's judged to be too corroded or unsafe). You and your electrician need to decide …

➤ If you can add any loads to your existing system.

➤ The compatibility of your current wiring with a new service panel.

➤ Whether your system is safe given its current usage.

➤ The practicality of completely rewiring your house.

➤ The cost versus the benefits.

Electrical Elaboration

If you decide to rewire your house and update your electrical system, remember one cardinal rule about electricians: They don't do wall repair! Electricians will cut into walls and ceilings, but you or another contractor will have to repair the plaster or plasterboard. The best electricians make the fewest possible openings to pull wires.

Do You Need to Replace?

Installing a new 200-amp service panel in an older, two-story house, replacing all the existing circuits, and adding new ones to bring the entire system up to code—all this is an expensive proposition. The service alone can cost roughly $1,800 to $2,000. You can easily spend four times that amount wiring the house, depending on its size and the complexity of the new system. I hesitate to quote figures because every house is different, as are local labor rates, but an electrician can give you a ballpark figure, which is subject to change when an actual estimate is drawn up.

Some people will replace and upgrade just to be on the safe side, while others *should* replace and upgrade. The following are signs that you should consider changing your electrical system:

➤ An undersized service (60 to 100 amps for a large, two-story, all-electric house)

➤ An insufficient number of circuits

➤ Too few receptacles and switch-controlled lights

➤ A lack of GFCIs

➤ Overloaded circuits with fuses that burn out regularly

➤ Frayed or deteriorated insulation on your current wiring

Too often, homeowners ignore the basic mechanics of a house (electrical, plumbing, and heating/air conditioning) when remodeling and pay too much attention to aesthetics, such as cabinetry, painting, and floor finishes. It's easy to understand why: These are characteristics we see day after day. No one sees the new, modern, sheathed cable running through the walls to equally new grounded outlets, all of which are nicely distributed on properly sized circuits. I can't begin to count the number of older homes I've been in that were beautifully redecorated but still only had one or two receptacles—and old receptacles at that—per room. There's no excuse for living in an underpowered house as we begin the twenty-first century.

Jump Up to 200 Amps

If you've got a small service, say up to 100 amps with a fuse box, in a two-story house and you're thinking of upgrading to circuit breakers, replace it with a 200-amp panel. Don't argue about it or debate its need, just do it. The last house we owned was a 1924, three-bedroom, one-story bungalow on a street of one-story bungalows. You could easily maintain that it only needed a 125-amp service, but a funny thing happened as Seattle real-estate prices headed toward the stratosphere: These one-story homes started becoming two stories. Owners decided it was more cost-effective to stay put and add on rather than to move. Five of our former neighbors on one block did this very remodeling to their homes. My point is that you cannot predict future needs. Given the relatively small cost difference between a 125-amp service and a 200-amp panel, there's no point in installing the smaller service in a two-story house.

Bright Idea

It's best to avoid disturbing elaborate plasterwork in older homes when rewiring. Carefully removing baseboards enables you to cut holes in your walls where they will go unnoticed while still allowing you to drill through wall studs and place your receptacles where you want them.

Positively Shocking

Even new electrical equipment can be potentially hazardous. In April 1999, the Consumer Product Safety Commission (CPSC), which can be found at www.cpsc.gov, announced a recall of more than 12,000 service entrance devices from a major manufacturer. These boxes contained the electric meter and breaker panel and were reported to have potential electrical arcing problems after one reported incident during a residential installation. Always have suspect service panels inspected by an electrician.

One exception to my 200-amp rule is the presence of gas appliances. Our first house, for example, was adequately serviced by a 125-amp panel because we had a gas furnace, water heater, stove, and clothes dryer. As a result, the biggest single electrical loads in the house were the refrigerator and the washing machine, neither of which were huge draws on the system.

Good Wire, Bad Wiring

An original service of knob-and-tube wiring, if properly installed, is and can remain a safe electrical system. Left alone, it would satisfy the electrical demand for which it was designed without any problems. When it isn't left alone, or when the loads increase and stretch the system's capacity, the problems and hazards begin. Add-ons are pretty easy to spot, especially those done by homeowners. Dead giveaways include ...

➤ Sloppy installation.

➤ Loose, unsecured wires.

➤ Wires running across the edge of floor joist rather than passing through them in unfinished basements.

➤ Improper taping at connections through the use of unapproved materials such as masking or adhesive tape.

➤ Mixing two different gauges of wire on the same circuit.

Other issues include worn and frayed insulation and brittle wire ends where they are attached to loads or switches. This problem is exacerbated as the wire ends get bent and unbent when switches, receptacles, and light fixtures are replaced over the years due to general wear and tear or the desire for something new.

When New and Old Collide

Old electrical systems can be safely added onto if the following considerations are followed:

➤ You have room in your service panel or fuse box for additional circuits.

➤ Individual existing circuits aren't fully utilized and can carry an additional load.

➤ The wiring and insulation of these circuits are intact and not worn.

➤ You properly join new wire to old.

Electrical Elaboration

Humans aren't the only ones who cause problems with old wiring—think rats! These rodents must gnaw to keep their teeth filed down, and electrical wire makes a dandy chew toy. Rats live inside insulation near hot-water heaters and furnaces. Mice aren't much better, settling down inside large appliances such as dryers or stoves and chewing on the wiring. Orkin Pest Control suggests that many fires of "unknown cause" might have been caused by rodents chewing through electrical wires. Cable television and phone wires also are targets. In New York City, rats have been known to gnaw through concrete—but that's New York for you.

You can't see inside your walls and observe how every receptacle, switch, and light has been wired, but you can get an idea if an addition has been made to the original system. Look for the following telltale signs:

➤ A receptacle or switch style that doesn't match the others in your house (The cover plates also might be a different style.)

➤ A fixture located in an odd place such as a crawl space

➤ A receptacle that isn't as evenly spaced as others

➤ A receptacle that is cut into a plaster wall while all the others are cut into the baseboards—a common feature in turn-of-the-century homes

➤ Any switch or receptacle that is surface-mounted on a wall rather than cut into the wall

Bright Idea

Replacing your current system offers the opportunity to relocate your service panel to a safer or more convenient location. It also might be a fitting time to bury your service entrance conductors or supply leads (the wires from the utility pole to your service panel) underground rather than having them suspended overhead where they are exposed to the weather and possible falling tree branches during windy weather.

Do you have unfinished attic or basement space? These are prime areas for added fixtures to be tied into existing wiring and should be checked thoroughly.

Location, Location

One consideration with old systems is the location of your fuse box. It's often located outside the rear or kitchen door, which isn't too convenient if you have to change a fuse on a cold winter night. Modern service panels with circuit breakers are located inside and are protected from the weather. Locations vary depending on the type and cost of the house, but typically the panels are found in basements, garages, or utility rooms.

Your Checklist

By now, you've got a better idea how fuses, circuit breakers, circuits, wires, and fixtures all come together as your house's electrical system. The older your house, the more important this knowledge will be as you evaluate your system's safety and consider future remodeling. The following checklist summarizes what you should be looking for in your system.

Electrical Checklist

Type of service
- ❏ Fuses
- ❏ Circuit breakers

Size of service
- ❏ 60 amps
- ❏ 100 amps
- ❏ 125 amps
- ❏ 150 amps
- ❏ 200 amps

Type of wiring
- ❏ Knob-and-tube
- ❏ BX
- ❏ Ungrounded cable
- ❏ Grounded cable

Dedicated circuits
- ❏ Major appliances
- ❏ Refrigerator
- ❏ Workshop
- ❏ Office computer

Electrical Checklist

GFCIs
- ❏ Kitchen
- ❏ Bathroom
- ❏ Basement
- ❏ Garage
- ❏ Outside

Sufficient circuits?
- ❏ Yes
- ❏ No

Do any circuits regularly trip?
- ❏ Yes
- ❏ No

Number of receptacles
- ___ Kitchen
- ___ Bathroom
- ___ Bedrooms
- ___ Living room
- ___ Dining room
- ___ Hallways

Wall switches for all lights?
- ❏ Yes
- ❏ No

Are all outlets polarized?
- ❏ Yes
- ❏ No

Are all outlets grounded?
- ❏ Yes
- ❏ No

Has the system been altered?
- ❏ Yes
- ❏ No

Does the work look professional?
- ❏ Yes
- ❏ No

Has your system been evaluated by an electrician?
- ❏ Yes
- ❏ No

Remember to follow all the safety recommendations in this book, especially if you have an old system.

The Least You Need to Know

➤ Branch circuits ensure an even and safe distribution of electricity throughout your house.

➤ A circuit map will tell you exactly which fuse or circuit breaker controls each load in your system.

➤ GFCIs should be installed in every kitchen and bathroom and next to any laundry or bar sinks, especially if you have an old electrical system.

➤ The main problem with old wiring is questionable work that was done *after* the system was installed.

➤ If you're installing a new service panel with circuit breakers, think larger rather than a minimal service size.

Chapter 6

When You Buy a House

In This Chapter

➤ House plans and inspections

➤ Some simple tests

➤ Aluminum wiring problems and issues

➤ Checking outdoor wiring

➤ Testing everything

➤ Keeping upgrade costs in mind

Your electrical inspection and checklist don't just apply to your current home but also to a prospective residence. Despite the diligence of inspectors, they still can make errors when checking and approving newly constructed homes and remodeled additions. Any home purchase, regardless of the age of the house, should include a general inspection and an electrical inspection. This inspection will be more exacting than one in your own home because you will have no familiarity with a house that's new to you, whereas you already know that your kitchen lights start flickering an SOS in Morse code every time you turn on your food processor and that they need to be corrected.

An inspection will tell you more than simply the condition of the system. It will enable you to consider changes and additions and their possible costs before you buy. These factors can affect your purchase negotiations, especially if major work is called for such as installing a new service panel.

New homes, both those of your own design and those under construction by developers, deserve special attention with regard to your electrical needs. The planning and construction phases are certainly the ideal times to wire for as many light fixtures, receptacles, dedicated circuits for your computers, and specialized cable for phones and media as you desire. If you're not sure about future TV or phone locations, wire every room for them and don't worry about it. The labor cost to install wiring is far cheaper when the walls are open to the framing studs than after they're finished.

Grab your checklist from Chapter 5, "More Wall Talk," and your Sunday newspaper's real-estate section and hit those open houses!

Caveat Emptor or Buyer Beware!

Whether it's new, old, or in its teenage years, you're taking a certain chance when you buy a house. All the warranties and assurances in the world won't prevent leaks, squeaks, and lawsuits over roofing or siding materials, for example. I can speak from experience on the last one, because our house—purchased new in 1994—features the infamous Louisiana-Pacific LP Siding, the subject of a massive class-action suit and a multi-million-dollar settlement. The same scrutiny applies to electrical work.

You have the law on your side with seller disclosure *Form 17* and builder warranties, but you want to preclude any problems after you move in by ferreting them out before signing the final papers. An inspection and a disclosure form keep a seller honest and can bring up unseen problems that were unknown even to the seller.

Ask an Electrician

Form 17 is a legal document that sellers are required to fill out unless the buyer takes the property as-is. The seller must disclose any known material issues that will affect the purchaser including structural and mechanical issues. The seller must disclose any remodeling work and state *whether the work was or was not done under a permit and inspection.*

Who Does the Inspection?

Professional house inspectors are very much a mixed bag. Some are extremely thorough and produce very detailed reports; others are less-impressive and depend too much on checklists and filling in the blanks. The very best person to inspect your electrical system is a licensed electrician, but you don't want to be dragging one along every time you look at a prospective house. Save that for final candidates. Meanwhile, you can do a preliminary inspection yourself.

The Preliminaries

Following the checklist in Chapter 5 will give you an overall view of an electrical system. Note whether the house has fuses or circuit breakers and the size of the service. Look for an overall impression of the condition of the wiring and the number of loads. A sure sign of a shortage of receptacles is the presence of multiple-outlet plug strips or multiple outlet plug-ins, especially in the kitchen.

Are there enough ceiling lights? Do rooms seem too dark? Keep in mind that every room should have a switch-controlled light fixture. This *doesn't* mean the fixture has to be installed in the ceiling or on the wall. A switch-controlled receptacle, a common feature in new homes, meets this code requirement because a floor or table lamp can be plugged into the receptacle and turned on from a wall switch.

Testing! Testing!

With a couple of simple testing devices, you can check for wiring problems including ...

➤ Whether power is present at a device or fixture.

➤ Grounding continuity.

➤ Defective receptacles, switches, and fixtures.

➤ Whether outlets are properly wired.

A voltage tester should be in every electrical do-it-yourselfer's toolbox. Consisting of two probes connected at a plastic housing that contains a small neon bulb, a voltage tester lights up when it detects an electrical current. It also can detect which incoming wire is the hot wire and the presence of a grounding conductor.

Positively Shocking

When examining a fuse box, note the number of 15-amp circuits. If you don't see any, someone probably messed with the box and installed 20-amp fuses instead—never a recommended practice. Lighting circuits almost always are 15 amps.

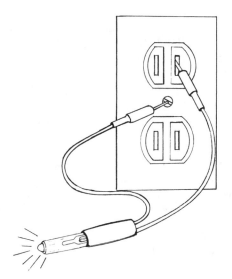

An inexpensive voltage tester.

The probes of the tester either are inserted into a receptacle's slots or are held against the terminal screws that secure the wire to the receptacle. If the probes do not detect a current when inside the slots but do detect one when held against the terminal screws, this indicates that the receptacle itself needs replacement. If there's no current at the screws, there's a problem with the circuit.

Note: With back-wired devices, the probes are inserted in the slots next to the wires. The receptacle will have to be removed from the box. To safely remove the receptacle, turn the power off at the panel first, then remove the screws securing the receptacle, pull it out, and, finally, test it.

Electrical Elaboration

There are two different means of attaching wires to receptacles and switches: back wiring and side wiring. The hot and neutral wires can either be inserted into two push-in terminals (also called terminal apertures or wire wells) on the back of the device (back-wired) or be secured with terminal screws on the sides (side-wired). Side wiring is considered by some electricians to be more secure, and it makes the wires accessible for testing. Back wiring is a faster installation, however. Test a back-wired device by inserting the probes into the release slots (which are next to the push-in terminals).

The same holds true for switches and fixtures. There are two tests for electrical switches.

After removing the cover plate, be sure the switch is in the "Off" position. Place one probe on the metal box that holds the switch; if the box is plastic or nonmetallic, place the probe on the white or neutral wire. You'll have to remove the switch from the box in order to reach the neutral wire; be sure to turn the power off first and then turn it on again for your test. Place the second probe on each of the black wires, or on the terminal screws holding the wires if they're side wired. One of them—the line side or black wire supplying power from the circuit—should light up. If neither of them does, there is a problem with the circuit.

After you find the line side, turn the switch to "On" and place the probe on the other black wire, which is the load side, while keeping the other probe on the neutral wire. In the "On" position, the switch completes the circuit and the load side carries power to the light fixture. If the tester doesn't light up, the switch is faulty and needs to be replaced.

You test a fixture by holding the probes against each of the terminal screws with the light switch "On." If your test shows a current, but the light isn't working, you need either a new fixture or new light bulbs.

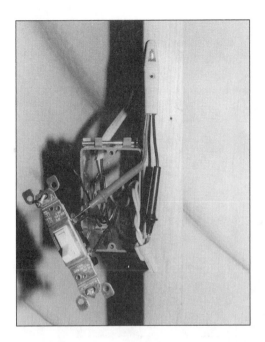

How to test a switch: line side terminal.

How to test a switch: load side terminal.

The Other Tool You Need

Second to the voltage tester in your box of tricks is a receptacle analyzer. This is a plug-in tester that analyzes different wiring problems after being plugged into a live circuit. Three lights on the bottom of the terminal light up in different combinations to show you what the tester detects. A printed guide on the analyzer points out the various electrical foibles a receptacle can suffer from.

A receptacle analyzer.

Bright Idea

You should change the batteries in your smoke detectors once a year to make sure they'll still be working in case of an electrical outage. Mark your calendar using a yearly event such as your birthday (or maybe October 8, the anniversary of the Great Chicago Fire).

Smoke Alarms

The best smoke-alarm systems are connected to your house's wiring and have a battery backup. A stand-alone, battery-powered alarm is better than nothing at all, but a wired/battery system is preferred. Check the house for smoke detectors and then check the detectors themselves. Press the test button on each smoke detector. If you don't hear the annoying screech designed to just about wake the dead, you have a dead battery, a dead circuit, or a dead detector.

Aluminum Wiring

Human beings always are looking to substitute new, less-expensive versions of successful products that have been tried and true for years. Sometimes this works well (a cheap, hand-held hair dryer versus a full-size, hair salon, sit-in-the-chair-and-put-your-head-inside hair dryer); sometimes it doesn't (Yugos and Vegas versus most other automobiles). In the electrical world, the use of aluminum wire for running branch circuits falls into the latter category of substitutes gone bad.

Aluminum wire was installed in at least 1.5 million homes between 1965 and 1973. The material cost was as little as 50 percent of the price of copper wire, which made it a hit with homebuilders, even if it ended up being a false bargain for homebuyers. Unforeseen problems with the connections of the wire to devices lead to it being labeled a potential fire hazard and ultimately banned from most residential use.

Although I could find no figures as to the actual number of homes that burned down due to electrical fires from aluminum wiring, there were enough to initiate studies, accusations, lawsuits, and not-so-veiled warnings regarding its use. The problem wasn't immediately apparent because aluminum-wired circuits can take years to reach a failure point while still remaining functional. According to Dr. Jesse Aronstein (in his report "Reducing the Fire Hazard in Aluminum-Wired Homes," prepared for the Electrical Safety Conference-Electrical Fires at the University of Wisconsin-Extension in March 1982 and revised May 10, 1996), a seemingly indefatigable researcher in this area, "The probability of an aluminum-wired connection overheating in a home varies considerably according to the types of connections, the installation methods used, and the circuit usage, along with many other factors. Without detailed knowledge of the installation in a particular home, it is not possible to provide specific advice on corrective measures."

Is aluminum wiring a red flag in your house-purchasing adventure? Yes, but there are ways to deal with it intelligently.

What's the Story?

According to the U.S. Consumer Product Safety Commission (CPSC), problems with aluminum wiring manufactured prior to 1972 include …

➤ Expansion and contraction of wires.

➤ Easily damaged during installation, because it's a soft metal.

➤ Corrosion.

Aluminum wiring heats up more easily than copper wire from electrical currents passing through it because it has a higher resistance. As a result, aluminum wire must be one gauge size larger for a given circuit than if copper were used. Thus, a 15-amp circuit could use No.14 copper wire but would require No.12 aluminum.

As a conductor, aluminum heats up when a current passes through it. Like any heated wire, it expands and contracts as it heats and cools, but

Positively Shocking

CPSC research indicates that homes containing aluminum wire manufactured before 1972 are 55 times more likely to have one or more connections reach "fire-hazard condition" than homes wired with copper. These conditions are defined as receptacle cover-plate screws reaching a temperature of 300°F, the presence of sparking, or charred material around the receptacle. Wiring produced after 1972 solved some, but not most, of the material's failings.

aluminum is damaged more than copper by this cycle of temperature changes. Adding to this problem are the connections (or terminations) at devices and fixtures. Aluminum tends to oxidize when it comes in contact with some other types of metals—the same ones that often compose the termination material (such as brass terminal screws). Now we've got a metal that's already touchy about heating and cooling, and it's also corroding and offering even more resistance to the current. The corrosion adds to aluminum's natural resistance, making that resistance even worse. As a result ...

➤ The connections deteriorate and loosen at the terminals.

➤ There is arcing or a discharge of electricity across the gap between the end of the wire and the terminal.

➤ There is possibly enough heat to melt the insulation and cause a fire.

Aluminum wiring can easily be damaged because it's so soft. If a piece gets nicked while the insulation is being stripped during installation, the nicked area is weakened and can deteriorate faster than the rest of the wire as it heats up. So much for that 50 percent savings in material cost when this stuff was installed!

What You Should Look For

Aluminum wiring primarily affects housing built from 1965 to 1973, but it also can be present in additions and remodeled sections of older homes if the work was done during these years. The first thing to look for during your snooping is the word "aluminum" printed or embossed on the plastic sheathing of the electrical cable. If the cable isn't observable in an attic space or a basement ceiling, look in the service panel. Remove the cover plates and look at individual switches and receptacles to see whether the wire ends are copper- or silver-colored.

Electrical Elaboration

Despite its dangerous reputation for branch circuits, aluminum wire is a good choice for conducting large currents such as the main service line to your house. It's lighter weight, so it's less affected by strong winds or added weight from winter ice. It's also less expensive than copper. Because there are fewer termination points (only those at your service head, where the service conductors enter your house, and at the service panel), they are tightly connected and easy to monitor.

Warning: Aluminum Wiring Ahead

Not every house with aluminum wiring is an automatic time bomb waiting to burst into flames like a vampire caught outside at dawn. The warning signs used to check for overheating problems and loose connections include ...

➤ Warm cover plates.

➤ Flickering lights.

➤ No power in a circuit.

➤ Sparks and arcing.

These signs are discounted by Dr. Aronstein, who states that they are not entirely reliable, do not always take place, or occur at a late stage in the failure of the wiring. He does, however, suggest turning the power off and examining individual devices and loads with a flashlight for the following:

➤ Charred or discolored plastic

➤ Back-wired devices

➤ Excessive tarnishing at the ends of the wires

➤ Damage to the thermoplastic sheathing including melting and bubbling

Some wiring might not show any signs of deterioration because a particular circuit might never have had enough loads on it to cause overheating. If you plug a portable heater in, however, the status quo might change rapidly. The best defense is to monitor these circuits if their usage changes.

Any of these defects is cause for action. Are they enough of a reason for you to walk away from a potential property purchase? No, because even though technology caused the problems, it also can resolve them.

Solutions

The obvious answer to any aluminum-wiring problems or potential problems is to replace it all with copper. How practical is this, however, if you have limited unfinished areas such as crawl spaces to run new wire or if you live in a condominium? You also would have to patch and repaint everywhere. Fortunately, there are other solutions.

Aluminum wiring is most dangerous at the connections and termination points. The accepted remedy is to use a short piece of copper wire (usually referred to as a "pigtail") to connect the aluminum wire to the switch, receptacle, or appliance after treating the exposed ends of the aluminum wire with antioxidizing paste (see the following figure). The best method, which is CPSC-approved as the only permanent repair, is done with the use of a special power crimping tool manufactured by AMP Incorporated (P.O. Box 3608, Harrisburg, PA 17105; 1-800-522-7652). This tool installs

Positively Shocking

When repairing aluminum circuit wiring, it's important that all devices be addressed and all wires be spliced and crimped rather than only the most noticeable problems. Every switch, receptacle, and fixture is a potential fire hazard and should be repaired by a licensed electrician.

a metal sleeve called a COP/ALUM parallel splice connector. This handy tool, which you'll never find in a hardware store, permanently attaches a piece of copper wire to the existing aluminum wire along with the sleeve. The connection is then covered with heat-shrunk insulation. This work should be done by professional electricians, not homeowners, so forget about trying it yourself with a cheaper crimping tool or another means of connecting the wires together. This is a specialized procedure that requires training provided by the manufacturer.

Note that some electrical boxes—the enclosures around switches, receptacles, and other devices—might not be big enough to house the additional pig-tailed connectors and wire. It might be necessary to replace the box with a larger size, which will involve cutting into the wall and doing some patchwork to the plaster or plasterboard.

A typical pigtail with NMB copper cable.

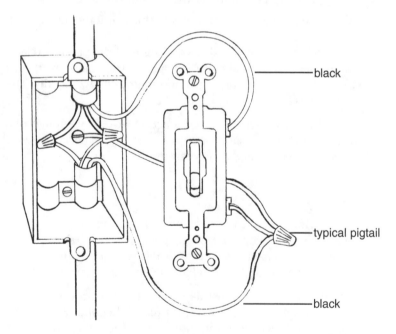

black

typical pigtail

black

A partial repair can be made by replacing all standard outlets and switches with UL-approved devices marked "CU/AL," which indicates that they can be used with either copper or aluminum wiring. The CPSC does not recommend these devices as a complete repair. Under no circumstances should any device connected to aluminum wire be back-wired. (If there are problems, you won't see them.)

The consumer booklet "Repairing Aluminum Wiring" (Publication 516) is available by writing the CPSC at U.S. Consumer Product Safety Commission, Washington, DC 20207.

The Great Outdoors

Look closely at any outdoor lighting and receptacles. Bad wiring practices outdoors are especially unsafe because of the exposure to weather, especially moisture. All outdoor receptacles should be GFCI protected and should be enclosed by weatherproof boxes. Electrical cable must either be buried in plastic or rigid metal conduit or be the UF (underground feeder) type, which has a very heavy plastic sheathing.

Cable rated for outdoor use must be buried at specified depths (see Chapter 21, "The Great Outdoors"). Anything buried too shallow for its particular type and circumstances is in violation of the code. Digging is a lot of work. I know one homeowner who did only a minimal burying, and the cable resurfaced from time to time during gardening chores. At another house, the roots of a growing tree gradually yanked a shallowly buried (and poorly located) cable to the surface.

Old, corroded switches and receptacles with broken covers are a sure sign of needed electrical repairs. Check the service panel or fuse box and see if the outdoor fixtures are running on their own circuit. You might find that they are tied into another branch circuit inside the house, which isn't a terrific idea.

Garages are another center for do-it-yourself add-ons run amok. The wiring here should be equal to any circuit inside your house and should not be a disarray of wires and extension cords stapled to the walls. Be sure to take note of any questionable work.

Bright Idea

Digging trenches for outdoor wiring is a lot of work, so combine it with another task such as bringing in a new gas line or service line for your plumbing. As long as you follow the local code requirements for maintaining a safe distance between pipes and wires, they all can share the same trench.

Check the power lines as well for obstructions or worn cables. These are the utility company's responsibility, but you need to report the problems so it can repair them.

Attic Insulation Problems

Modern nonmetallic electrical cable is tough stuff, and it doesn't require the open-air environment of older knob-and-tube wiring for heat dissipation. Even if you pack it up against insulation, it just laughs as long as cable-nibbling rodents don't take up residence. Knob-and-tube wiring, on the other hand, can have a dysfunctional relationship with insulation due to heat buildup. It isn't unusual to find knob-and-tube wiring buried under the blow-in type of insulation, so check your local code to see if this is allowed.

Older light fixtures also need clearance from insulation because they really build up heat, especially if they have more than one light bulb. Although newer zero-clearance fixtures are now sold, never assume they have been installed in an older house until you examine the fixture itself.

A Breath of Fresh Air

Modern homes are tightly insulated, but they have plenty of mechanized ventilation in the form of kitchen and bathroom fans as well as whole-house fans on timers. These forms of ventilation try to ensure a regular exchange of fresh air, which is particularly helpful the first year after construction because carpets, plastic laminates, and paints give off all kinds of fun fumes, thanks to the wonderful world of chemistry. An old house is naturally ventilated through gaps in the windows, doors, and walls. Even so, you still should have fans in the kitchen and bathrooms. If these rooms remain unventilated, moisture hangs around and throws a party for its friends mildew and mold. This means more frequent cleaning and painting as well as a reminder that you burned the pancakes at breakfast because you can still smell them during dinner. Check for fans and make sure they're working. Small fans (those with a low *CFM* or *cubic feet per minute*), whose main activity is making noise instead of getting the moisture and odors out, are next to useless and should be noted for replacement.

Ask an Electrician

CFM (cubic feet per minute) is a measure of how much air a fan can move. A good fan can move—and remove—a lot of air quickly.

More Testing

Turn on all the lights in each room. Next turn on various appliances, even hair dryers in the bathroom, and watch the lights for flickering. A slight flicker isn't unusual, but a permanent dimming is a sign of an overloaded system. The same is true if an appliance repeatedly causes a circuit to trip.

If a switch doesn't appear to control anything when you flip it on, it can be a sign of an old disconnected load or circuit. A do-it-yourselfer might leave the switch in place, but an electrician would remove it and cover the box with a blank cover plate. You want to check for signs that work was done properly, not slipshod.

Some Final Points

Does the prospective home have a security system? If you want one, make a note regarding its installation and approximate cost. Modern systems have battery-powered motion detectors that don't require hard wiring, but the basic unit itself is wired, which means you need to supply power to it. This might not be a big deal, but it's one more thing to add to your electrical wish list.

Do all the rooms have heat, either from a central heating system or electric room heaters? Some older homes have ancient heating systems and might only have one heating vent in an upstairs landing rather than heating for each individual bedroom. This means you'll have the expense of supplying heat either with new ductwork (and probably a new furnace) or by running circuits for electric heating.

Electrical Elaboration

Check your local code while you consider installing individual electric room heaters. Some cities and municipalities don't like to increase electrical demand any more than they have to, and they might require you to upgrade your house to higher insulation standards, including insulated windows, if you install electric heat. They would much rather you use gas or oil heat, which does not affect their utility loads.

Old homes offer a certain charm and comfort that are difficult to duplicate in most modern housing. Leaded windows, wide oak entry doors, and crown molding running along the ceilings aren't common features in new construction. Neither are fuse boxes, tacky wiring repairs, and a shortage of receptacles and lights. Updating an old system will have to be figured into your budget (and ultimately your purchasing considerations) when you start house hunting. Throw in a new roof, plumbing, floor refinishing, and a kitchen addition, and Old World charm takes on an expanded definition—bring money!

A complete house inspection will make you a more knowledgeable homebuyer and will prepare you for the true cost of purchase.

The Least You Need to Know

➤ Always do a thorough electrical inspection of any prospective house purchase.

➤ A voltage tester and a receptacle analyzer are two inexpensive but indispensable tools for testing your circuits and electrical devices.

➤ Learning to live comfortably with aluminum wiring usually requires the services of an electrician.

➤ Kitchens and bathrooms need ventilation fans—usually the bigger the better.

Part 2

Safety, Tools, and Contractors

A modern electrical system should be both safe and convenient. All of its components, from the transformer on the utility pole to your imported espresso machine, are supposed to work together in quiet, Buddhist harmony as electrons zip through branch circuits. When you introduce bad karma in the form of overloaded circuits, wiring that doesn't meet code, or unsafe work practices, you can forget about reaching electrical enlightenment anytime soon. You need to maintain a safe environment in your day-to-day dealings with electricity and when doing your own work on the system.

There's a saying that you can tell the quality of a worker by the quality of the tools used on the job and the way they're stored and maintained. This is a little too simplistic (a Wall Street Journal *article, for instance, featured homeowners with a lot more money than skills buying tools that were clearly beyond their abilities), but good tools will make your job easier. Consider the value of your time when you work up a tool budget. How do you measure the frustration factor and the lost time from using an underpowered drill because it cost $30 less than a more appropriate one? You don't need the very best tools because you won't be doing electrical work for a living, but you also don't want to try and slide by using your kids' "Building Trades for Tots Toolbox."*

Safety has been an issue with electricity since the days of Thomas Edison and the first large electrification projects. It's still an issue despite the multitude of advances we've made since the first light bulb was invented. Every year, there are thousands of injuries and electrical fires in the United States, and you don't want to add to the statistics. I will emphasize over and over throughout this book that you need to work safely and follow the local code requirements. Electrical safety also applies to everything you plug in and turn on, so I'll discuss those items, too.

Chapter 7

Caution Signs and Safety Concerns

In This Chapter

➤ Skip the shocks

➤ Common electrical problems

➤ Safe work practices

➤ Fire, a major possibility

➤ The best practice is vigilance

At different times in our childhood, well-meaning parents, teachers, and other uninvited authority figures decided we needed a lesson on what they viewed as one of life's serious dangers. These lessons usually consisted of terse descriptions of the horrors and grave consequences that would befall us if we ignored the lessons, which most of us did. I can confidently say that I didn't go blind from sitting too close to the television, nor did I ever catch pneumonia from going outside with my coat unbuttoned in the dead of Ohio's winters.

Electricity is a different matter, however, and here *I* must become parental for your own good. (I'm sure that phrase brings back a few memories.) One wrong electrical move can result in injury or even death for you or your loved ones, not to mention fire damage to your home. No one is immune to accidents, including experienced electricians. As a homeowner and do-it-yourselfer, you should keep yourself safe and be even more cautious than a professional.

A brand-new electrical system doesn't give you license to abuse it or to test its limits just for the fun of it. Sticking your house key into a ground-fault circuit interrupter (GFCI) receptacle while grabbing hold of your bathroom sink's faucet isn't recommended, even if the GFCI should shut down when it detects your act of lunacy. There's always a chance it *won't* shut down because it's defective. Of course, if you've been following safe electrical practices, you would know this because you would test your GFCIs every month.

With a little common sense (another childhood admonition), you'll be able to safely inhabit and work on your electrical system and fill your home with lighting, receptacles, and multimedia features from top to bottom.

Positively Shocking

It's the salt in water that makes it so conductive. When the salts dissolve, they form electrically charged ions that attract the charges of an electric current. Even the purest water in a scientific laboratory will conduct a minute amount of current. If you're standing in water or a damp area, you're more likely to conduct current. Our own body salts, carried through the sweat glands, also make us more conductive.

Shocks Galore

Every time we use electricity or are near it, there's a chance we could get shocked. Sometimes the chance is remote, such as when turning on a living room ceiling light. Other times, the chance is very good, such as when plugging in a string of worn, patched, and taped-together Christmas tree lights left over from the days of the Harry Truman administration. We talk about shock and its big brother, electrocution, but what exactly are they? Why are they so hazardous to our health?

Electricity basically is lazy and is not always interested in staying on the straight and narrow path of an alternating current. Given the choice of making the return trip along the neutral wire or taking a shortcut, it will opt for the shortcut every time, even if it means traveling through your extremities. Electricity seeks the easiest path to the ground, and any available conductor—metal, water, you—will do the job. Because our bodies are 70 percent water, we make it easy for errant electricity to hitch a ride, and it does so without any hesitation. When our skin is dry, it blocks electricity pretty well but not when there's water around or the current is sizable. The size of the current and the duration of our exposure to it are the real health issues.

When You Can't Let Go

When an electrical current starts passing through your body, it doesn't take much for you to become energized and very attached to that current. The "can't let go" level (or freezing current) for adults is small, around 10 milliamps. Young children can get stuck at half that level. The path of the current is of critical importance as well. A hand-to-foot pathway will involve vital organs, especially the heart, and this can have serious consequences.

The following are some effects of an electrical shock:

➤ Knocking someone down or away from the source of the shock

➤ Respiration disruption

➤ Unconsciousness

➤ Muscle spasms

➤ Seizures

➤ Interrupting the heartbeat

➤ Severe burns

➤ Cardiac arrest

The longer the contact and the greater the current, the greater the injuries. A young adult in good health will be less affected by an electrical shock than a very young child or an elderly person, but you still don't want to take any chances.

If the current is great enough, third-degree body burns can result at the points of entry and exit. Burns damage and destroy the skin, further breaking down its resistance to the current.

How Much Can You Take?

Once again, we run into Ohm's Law of electrical resistance. In the case of a shock or an electrocution, the amount of current zapping its way through the body is determined by the following formula:

$I = E/R$

I = Electrical current

E = Voltage

R = Resistance of the body

Every body offers a different degree of resistance, but that doesn't mean you want to challenge the averages. The National Electric Code (NEC) figures five *milliamps* to be the safe upper limit of exposure for children and adults. Even at this level, you still can be injured by your reaction to the shock such as jumping back and tripping over the rim of a bathtub.

Ask an Electrician

A **milliamp** is a measure of electrical current. It's very small, equal to one thousandth of an amp or ampere. Its abbreviation is **ma.**

Timing Is Everything

The longer a victim is exposed to an electric current, the greater the chance of critical injury. In addition to burns, there is also the loss of muscular control, breathing difficulties as the chest contracts involuntarily, and ventricular fibrillation of the heart. This last effect comes up repeatedly in any discussion of severe electric shock. It refers to rapid, irregular heartbeats and equally irregular fluttering of the heart muscle.

It's one thing to have your heart skip a beat or two because you're head-over-heels in love, but it's quite another to have its pumping activity disrupted because of a faulty circuit. The former usually is a lot of fun, but the latter can do you in if it goes on for too long.

Know Your First Aid

Chances are, you'll never have to rescue anyone on the receiving end of a severe electric shock unless you work in certain construction specialties. Any of us can be caught in a lightning storm, but the chances of being struck by lightning are remote. Nevertheless, it's worth being prepared in the event of an unforeseen accident.

There are a few cardinal rules to remember when helping electrocution victims:

➤ Assume that the victim is still in contact with the current.

➤ Never touch the victim until you're certain the current has been shut off or the victim has been removed from the current. (Otherwise, you can be electrocuted, too.)

➤ If you can safely do so, shut off the power source at the fuse box or service panel. If it's more practical, pull the plug from the receptacle.

➤ Push or pull the person from the power source using something nonconductive such as a wooden broom, a rubber mat, or a plastic chair. Don't use anything made of metal.

➤ Never directly touch the source of the current.

➤ If the victim has stopped breathing, call 911 and begin CPR. (If the situation warrants, call 911 before attempting any rescue.)

In the event of a high-voltage electrocution such as an industrial situation or contact with power lines, do not attempt any direct rescue. Currents this strong can jump beyond the victim and hit the rescuer as well. Call the fire department and keep others at a safe distance.

Summer lightning storms bring their share of electrocutions as well. A bolt of lightning can carry millions of volts of electricity, far more than a misbehaving kitchen receptacle. When you assist a lightning victim, the current already has passed through, so you don't have to worry about being electrocuted as well (unless lightning decides to strike twice, of course).

The Source of the Problem

Most electrical injuries are preventable. They typically result from …

> ➤ Work that isn't done to code and is not inspected.

> ➤ Defective fixtures, devices, or appliances.

> ➤ Human error.

Positively Shocking

If you ever suddenly find yourself near a downed power line, keep yourself grounded by shuffling away to safety. Running will break contact with the ground and set you up for a shock.

The major purpose of the NEC and your local codes is to prevent injury and property damage from the use of electricity. Even when all the electrical ducks are lined up in a nice, neat row, human error or ignorance comes into play because we regularly ignore good safety practices. Grade yourself by taking our first *Complete Idiot's Guide* electrical quiz and see how you do.

Quiz #1

When you see a worn electrical cord on a lamp or an appliance, do you …

    A. Tell yourself it adds to the ambiance of your home?

    B. Wrap it with lots and lots of electrical tape?

    C. Replace it with a new, same-size cord?

When you turn on an electrical appliance, do you …

    A. Make sure your hands are dripping wet?

    B. Grab on to a water faucet for balance?

    C. Dry your hands and stand away from the sink?

Before cleaning the bread crumbs from your toaster, do you …

    A. Grab the cord and give it a yank?

    B. Not even bother to unplug it?

    C. Grasp the plug and pull?

Before working on an electrical circuit, do you …

A. Stick a screwdriver in a receptacle to check the current?

B. Make sure you're standing on a very wet surface?

C. Turn the power off at the breaker or fuse and stand on a dry board if the floor is damp?

To power the garage door opener in your detached garage, should you …

A. String a series of extension cords together and run them between your house and the garage, leaving them out in all kinds of weather?

B. Try to run a wire off your washing machine's receptacle?

C. Run a separate circuit with properly buried cable?

Okay, it's a trick quiz. If you answered anything other than "C" for any question, go back and start reading this book again. It isn't just major electrical work that requires vigilance; everyday stuff is dangerous, too. Problems can be prevented with even the simplest practices such as …

➤ Installing childproof safety caps on all receptacles.

➤ Avoiding overloading circuits with too many loads.

➤ Keeping ladders and tree branches away from power lines.

➤ Unplugging all small appliances when not in use.

➤ Turning off the power to any receptacle or switch that feels excessively warm to the touch. Follow up by having the circuit checked. (Note that dimmers are an exception: Often the heat created by the dimming function is dissipated through the screws holding the cover plate on.)

➤ Not tucking in an electric blanket or covering it with another blanket to avoid excess heat buildup.

➤ Keeping extension cord use to a minimum and never running cords under carpets or rugs.

➤ Replacing broken cover plates on switches and receptacles so wiring isn't exposed.

➤ Never leaving a lamp socket without a light bulb in it by replacing burned-out lamps immediately. Only replace them with lamps of the same wattage or lower, never exceeding the manufacturer's recommendation.

> **Electrical Elaboration**
>
> Some electrical engineers have gone into the consulting business as forensic engineers. Their services are regularly used in lawsuits involving electrical mishaps. Forensic engineers investigate electrocutions, electrical failures, and pain and suffering as the result of electrical problems. They also conduct accident reconstruction. They testify and consult on anything from problem infant monitors that injure a child to massive industrial accidents.

Safe Work Practices

The mundane world of GFCIs and using the proper-size light bulb in your light fixtures just scratches the surface of electrical safety. Its importance is heightened when you do any repairs or alterations to your system. It's like the difference between swimming in the shallow end of the pool and jumping into the deep end with the big kids. Unlike the big kids, who literally sink or swim based on their skill level, you can stack the deck in your favor with a few preemptive moves.

The number-one, top-of-the-list safety rule in electrical work is this: Make sure the current is off, and if it isn't, shut it off! After the power is off, any fumbling with the wires will be a forgiving experience instead of a highly charged one. An inexpensive voltage tester can tell you in seconds whether a current is hot.

Plugging a tool or a light into a receptacle is *not* an adequate way to test whether a circuit is on. The receptacle might be defective. Always test the conducting wires themselves with your voltage tester.

Turn It Off!

Shut the power off by either removing the fuse or flipping the circuit breaker. After the power is off, *hang a sign on the service panel stating that you are working and that the power should remain off!* You don't want someone flipping the breaker on thinking that it tripped.

> **Positively Shocking**
>
> Remove any rings or jewelry before doing any electrical work. These are dandy little conductors when they come in contact with a live circuit. Can't get your wedding ring off? Wrapping it with multiple layers of electrical tape will help, but this is no guarantee against shock. Wearing a work glove will give you additional insulation, but it gives you less flexibility to do the work.

One Hand Behind Your Back

An electrical current needs to travel from point to point to complete a circuit. If you grab the end of a hot wire with one hand and touch a water pipe with the other, you provide the current with a path as it travels through you. The same is true if you're standing on wet ground; the current will travel toward your feet. For this reason, whenever you change a fuse or flip a circuit breaker, you should use only one hand. The other hand should be behind your back or in your pocket. In other words, your other hand should be away from the service panel or fuse box. You don't want it to accidentally come into contact with any metal surfaces in the box because that completes a pathway for the current.

Test, Test, Test

Even if you've turned off the power to a particular circuit, check every device and fixture along the way before you do any work on them. You should take this precaution in case a light switch or receptacle is wired into a different circuit than the one you're working on. Industrial and construction electrical accidents happen regularly because workers find equipment or circuits to be energized when they believed the power was shut off.

On Dry Ground

When it comes to electricity, dry is good and wet is bad (unless you're the U.S. Army trying to electrocute some prehistoric swamp creature in a cheesy, 1950s monster movie). Never stand in a puddle or on damp ground when working on your electrical system. Always find a dry piece of wood or another insulating material to stand on while working. If you must work during wet weather, wear thick-soled rubber boots. Better yet, wait for a dry day (always an iffy situation here in the Northwest, unfortunately).

Watch That Ladder

Metal ladders and overhead power lines are a bad combination. Every year, painters and tree trimmers learn this the hard way, resulting in injuries and electrocutions. Using power tools while working off a metal ladder also can be hazardous, especially during wet weather.

Electricians use wood or fiberglass stepladders and fiberglass extension ladders when they work. Fiberglass is nonconductive; wood also is an excellent insulator as long as it's dry. Most extension ladders aimed at the homeowner market are inexpensive metal ladders. If you're doing any serious overhead electrical work, however, a wood or fiberglass ladder is the better choice.

Tool Health

Modern tools either are *double-insulated* or come with a ground pin in a three-pronged plug. Power tools greatly speed up just about any job, but you can't take them for granted. Tools with frayed cords, cracked casings, or incidents of sparking should be repaired or replaced. This is especially important with heavier-duty tools if they have metal casings. The casings can become energized if there's a short in the wiring.

Vigilance, as always, pays. Inspect your tools before starting a job. It's far easier—and safer—to catch and tape a small tear in a drill's power cord than to chance an injury.

A Lesson from Your Kids

As adults, we're supposed to exude maturity and responsibility and set an example for our children. We try to make sure they're fed, warm, and in bed on time. We don't always apply this same concern to ourselves, however, and this can be dangerous when we're working on our homes.

If you're cold, hungry, or tired, you could start making mistakes, so pay attention to your comfort level. Shaking hands, a growling stomach, and fluttering eyelids are signposts on the road telling you to pull over, put your tools down, have some lunch, and maybe catch a quick nap. We're viewed as a sleep-deprived nation, and there are plenty of accident statistics to affirm this view. You're not going to save any time or keep to a schedule if you have to redo some of your work later because it's faulty.

Speaking of Kids ...

Children also use electricity and need to use it safely. You, as a parent (or an adult friend), need to instruct them about the hazards of yanking electrical cords out of receptacles instead of holding the plug and pulling, using a hair dryer near water or with wet hands even if you have a GFCI in the bathroom, and sticking pens into receptacles to watch them melt. In addition to your always-welcome lecturing, a number of audiovisual helpers are available that could be shown at your children's school.

Positively Shocking

A flooded basement can be dangerous if the water comes into contact with any wiring, extension cords, or appliances. Don't enter your basement to start pumping the water out before the power is shut off. If you must go in, wear thick–soled rubber boots or call your local utility for assistance.

Ask an Electrician

A **double-insulated** tool doesn't have a ground pin in the plug. Instead, the wire conductor is surrounded by additional nonconductive material such as plastic. This does not guarantee against shock in the event of frayed wiring or damage to the tool. Metal casings also can be lined with nonconductive material.

Some of these films include ...

➤ *I'm No Fool with Electricity,* by Disney Educational Productions. This film, according to its catalog description, somewhat implausibly shows Pinocchio and Geppetto exploring electrical safety both indoors and outdoors. Because he was made from wood, Pinocchio has the built-in advantage of being an insulator instead of an electrical conductor, at least as long as he remains dry.

➤ *Electrical Safety from A to Zap,* from Perennial Education, Inc. In this film, a mouse shows a cat how to use electricity safely, their lack of opposable thumbs notwithstanding.

➤ *Play It Safe from HECO.* This video features two children who learn safe practices around electricity. The film's big plus so far in our list of audiovisuals is the fact that it features human beings who actually *do* use electricity.

➤ *The Electric Dreams of Thomas Edison: A Guide to Indoor Electrical Safety/A Guide to Outdoor Electrical Safety,* produced by the Southern California Edison utility company. In this film, students defy all the rules of logic and physics by somehow communicating with the long-dead Thomas Edison, who informs them about grounding, insulators, and conductors. They also look for outdoor electrical hazards.

➤ *Zap Rap,* from Pacific Learning Systems, Inc. Sure to appeal to the contemporary youngster, this film uses rap-style language to convey the wonders and dangers of electricity. As with most attempts to maintain students' interest through the use of entertainment as a teaching tool, you might give your kids a quiz to see if they learned anything at all about electricity other than a few tunes.

➤ *Fire in the Kitchen,* from Film Communicators. This video is aimed at grades 7 through 12. There are no wooden puppets or rappin' electrons here. This video teaches kitchen safety, including proper use of a microwave oven.

➤ *Our Invisible Friend—Electricity.* This 17-minute feature from Marcom Marketing Group also was made for grades 7 through 12. One has to wonder, of course, if any video made for seventh graders could be even remotely interesting to high school seniors.

➤ *Safety at Home: Electricity,* from AIMS Media. Geared for grades 9 through 12, a utility inspector shows careless use of electricity and the inevitable results. The inevitable results in grades 9 through 12 will be hooting and applause as actors are shocked and fried while doing things with appliances, receptacles, and plugs that a three-year-old wouldn't consider doing.

If my experience with audiovisual presentations when I was in school is still typical of students today (some things really don't change), I'd suggest that you take your children's electrical education into your own hands. Take them around the house, show them how the circuit breakers and GFCIs work, even show them how to properly insert and remove a plug from a receptacle. If they're old enough, turn the power off to a circuit, remove a switch or receptacle, and show them how it's wired. By the time they start getting bored, you'll have gotten the basics across.

Some Statistics

Mark Twain once said that there are three kinds of lies: lies, damn lies, and statistics. Trying to track down accurate figures about residential electrical fires produced quite a range of numbers. Everyone from the U.S. Consumer Product Safety Commission to various fire marshals across the country has a different figure to get the same point across: Misuse of electricity is a bad idea with sometimes incendiary results.

Based on my reading, the following figures are well inside the ballpark when it comes to fires caused by electrical problems:

➤ Approximately 45,000 to 50,000 fires each year occur in homes because of faulty wiring, appliances, and extension cords.

➤ The National Center for Health estimates that approximately 760 electrocutions take place from all causes each year including 310 occurrences involving consumer products.

➤ More than 3,000 children under the age of 10 are treated in emergency rooms each year after inserting objects into electrical receptacles. Another 3,000 are treated for injuries associated with extension cords.

➤ According to the CPSC, plugs and cords are involved in close to 20 percent of all residential electrical fires each year.

➤ Electrical fires kill hundreds of people in their homes every year, injure thousands more, and destroy hundreds of millions of dollars in property.

➤ December is the most dangerous month, electrically speaking, because of holiday lighting and portable-heater use.

➤ Older homes are more likely to have a fire than homes built in the last 20 years.

Some figures overlap, some are subject to various interpretations, and others, such as figures regarding electrocutions, may include industrial accidents. Sweeping all these distinctions aside, you don't want to be part of anyone's statistical table unless it's for happy, content homeowners with up-to-date electrical systems, cheerful children, and pets who do their business outside when nature calls.

Every holiday season, just about everyone's local news shows photos or a film clip of a family dispossessed and standing out in the cold because of a faulty space heater. It shouldn't happen to you if you follow the usual rules:

➤ Don't overload your circuits.

➤ Never replace a fuse or a circuit breaker with one of a greater amp rating.

➤ Don't run electrical cords under rugs or carpets.

➤ Carefully examine any wattage-hungry appliance, such as a portable heater, before using it, and then carefully monitor it while it's in use.

Bright Idea

Ontario Hydro (a Canadian power company) promotes the month of May as Electrical Safety Month, reminding people to keep a "safety first" attitude during outdoor activities. The company's tips include avoiding kite-flying near power lines, inspecting electric lawn mowers and power tools before each use, and avoiding mowing wet grass with an electric mower. Call the power company before you dig, it warns, so you don't hit any wires.

Electro Kindling

It's bad enough that a faulty electrical connection overheats inside a small appliance or device. What might even be worse is the fact that the wiring is surrounded by flammable materials, often plastic casings that have replaced the metal casings from years ago. Dr. Jesse Aronstein (of aluminum wire repute) came up with the term "electro kindling" to describe the material that ignites and burns after the failure of an electrical connection. A plastic toaster will be only too happy to burn if there's a wiring problem, while a metal one will just hang tight until you unplug it or until the wiring fries to a crisp. There have even been reports of multiple-outlet strips, icemakers, and plastic thermostats failing and subsequently igniting.

Such is the price of a society that embraces plastic in all its forms. The possibility of electro kindling should reinforce the practice of unplugging your small appliances when they're not in use.

New Service Doesn't Let You Off the Hook

I'm a big believer in new electrical services, but that doesn't mean you can ignore them completely, especially if they're tied into older, existing wiring. Breakers should trip if a circuit is overloaded, and GFCIs should shut down in the event of a ground fault, but there's always the chance you have a piece of defective equipment. Unlikely? Sure, but it's certainly possible.

Test your GFCIs monthly and pay attention to outlet and switch cover plates that seem too warm to the touch. If you ever smell anything burning around a receptacle, and it isn't an individual appliance or load, shut down the circuit immediately and call an electrician. It's worth the price of a service call for your peace of mind.

Electrical Elaboration

According to CFRA News Radio's Web site, a fire at the home of Dallas Cowboys quarterback Troy Aikman in March 1998 was attributed to faulty wiring. This wouldn't be an exceptional news item except for the fact that the $3.2 million home, which took more than two years to build, was brand new! The three-alarm fire at the Plano, Texas, home caused almost $200,000 in damages to the attic and the garage. It took 50 firefighters to extinguish the fire in the house where Aikman, who was away at the time, had lived for a total of three weeks.

More Information

For more information about electrical safety, contact one of the following agencies for printed material:

National Electrical Safety Foundation
1300 North 17th St., Suite 1847
Rosslyn, VA 22209
Phone: 703-841-3211
Fax: 703-841-3311

Send a self-addressed, business-size envelope with 55¢ postage for a copy of the *Home Electrical Safety Check* booklet.

The United States Fire Administration
Office of Fire Management Programs
16825 South Seton Ave.
Emmitsburg, MD 21727

The Least You Need to Know

➤ Preventing electrical shocks and electrocution should be a top priority with any electrical system.

➤ The most common causes of shock are bad wiring practices and human error.

➤ The number-one safety rule when working on an electrical system is to shut the power off.

➤ Children should be educated about electricity and how to use it safely.

Call Me Sparky

Wiring is a nice, logical process. You want to get power from point A to point B in the most efficient way possible. Running wire or "roping" a house is mostly a matter of drilling access holes through the house's framing (the wall studs, plates, and floor joist) and pulling electrical cable through those holes. *How* you carry out this nice, logical process is another matter altogether.

Like just any task in life, you can do your own electrical work the hard way or the easy way. The hard way means tearing open more walls than necessary, undoing and then redoing part of the job due to poor planning, and trying to drill holes, cut wire, and strip insulation with cheap tools. The easy way calls for planning and economizing your moves and using good tools to give you a better job and to move you through it faster. You won't be as fast as an experienced electrician, but you'll have the satisfaction of doing your own work and doing it well.

If you're going to be your own electrician, you need to take your role as seriously as a professional would. This means presenting any required plans to your building department when you take out a permit, knowing the code issues, using the right tools, and finding suitable suppliers for your materials and fixtures. You don't need to invest in the same level of equipment as an electrical contractor does. After all, you're not going to be making your living at this. You can, however, become a talented amateur whose work can be respected, even by a professional!

An Electrician's Mindset

An electrical contractor has the following goals in mind when bidding, planning, and actually doing a job:

➤ The job must meet the customer's requirements.

➤ The work must be safe, meet code, pass inspection, and be finished in a timely manner.

➤ The final result should be a satisfied customer and a profit for the contractor.

Your goals as a do-it-yourselfer shouldn't be any different. You want your work to be of satisfactory quality so you can live comfortably with it rather than going crazy every time you look at a crooked light switch or receptacle. It goes without saying that your work must pass inspection, but you also want to get it finished sometime before you reach retirement age. Money-wise, you want to realize a savings from doing your own work. This can be measured in different ways. Some people keep an exact accounting of their hours and assign an hourly rate to the job versus an electrician's labor bid. Others see it as using their off-hours productively, and anything they save is pure profit. However you measure your savings, you will have faced some challenges and learned from them, and you can't put a dollar value on that.

Bright Idea

Arrange to look at a house under construction so you can examine the electrical work, especially the service panel. This will give you a better idea of the minimum standards you'll have to meet when you do your own work.

By now, you'll have drawn up a plan, calculated the loads, and gotten your permits. A plan on paper will show you where to locate a receptacle or a light fixture, but it won't show you how to do so. For example, will you …

➤ Run your cable through the attic and drop it down between the wall studs?

➤ Consider using the basement or a crawl space for access?

➤ Use wire molding and run it along the surface of your baseboards?

➤ Remove some of the wood trim and cut into the walls behind it to avoid patching more noticeable sections of the walls?

A half-hour of forethought and planning can save you hours of patching. If there's more than one solution to the job, look at them all and decide on the best approach.

Think Before You Drill

Drilling an unnecessary hole or two through a wall stud or a floor joist isn't a big deal. Visitors will never see it, and you're not weakening your house's framing. This won't help you get the job done any faster, however, especially if you drill a whole series of holes in the wrong places. Drilling the wrong hole in plaster or drywall is another matter because it requires a repair. In the worst-case scenario, let's say you cut a four-inch-diameter hole in the ceiling with a *hole saw* only to discover it was in the wrong location. You would have to …

➤ Patch in new drywall or plaster.

➤ Coat the patch with finish plaster or drywall compound until it's smooth. (In the case of a textured surface, the texturing would have to be matched.)

➤ Prime and paint the patch.

➤ Possibly paint the entire ceiling to match the patched area.

What if you're not sure where you want to locate a light fixture? Attach some blue painter's masking tape (this type doesn't dry out as quickly as regular masking tape) to the proposed location and leave it up for a day or two. Apply the tape in roughly the same shape as the fixture. If it's a hanging fixture, also attach a string the same length as the chain or light cord. You might decide you don't want hanging lights, or you might want to relocate them. When you're satisfied you've found the right location, you can start cutting into the ceiling.

Minimum Damage, Minimum Repairs

Hole saws and other tools of the trade do more than make pretty holes for your electrical work. They also keep damage and subsequent repairs to a minimum. This is especially true when you're cutting through plaster and lath. Lath is the wood or metal backing that acts as a form for wet plaster. The plaster is forced into the lath where it eventually dries into a wall or ceiling. It's almost impossible to cut a clean hole through lath using hand tools. A hole saw or

Ask an Electrician

A **hole saw** is a circular saw attachment for electric drills. It is used to cut through walls and ceilings instead of using a hand tool, usually a **small keyhole saw.** A hole saw renders a smooth, neat opening for a round electrical box used for most light fixtures. It's important to keep the hole saw straight when in use so it doesn't bind and cause the drill to turn and possibly slip.

a reciprocating saw will do the job quickly and cleanly. Believe me, you'll never catch an electrician using hand tools when a power tool is the better choice.

You might think you have limited use for a hole saw, and you might not want to spend the money on one, but there's another way to look at it. A four-inch hole saw costs around $20. If you have to install five light fixtures, the hole saw is costing you $4 a light. It's also greatly reducing your labor time, and you can avoid the frustration of trying to cut a clean circle with hand tools. Besides, you can always find a future use for it, especially if you have more remodeling projects in mind.

Permits

I'm not going to pretend that a permit is taken out for every electrical job, even if the local building regulations call for it. It's tough to justify the time and expense to obtain a permit when you're only adding a single receptacle to a circuit that can easily support the addition. Nevertheless, I would be remiss as an author if I ever advocated anything less than playing by the rules, especially given the possibility of harm and damage from wayward electrical work. As a case in point, our own electrician recently was telling me about some receptacles added in a residence—by another electrician—without a permit, and none of them were grounded correctly.

Bright Idea

You can always discuss your plans with your local building department before you take out a permit. It might point out any deficiencies and guide you through the permit process. Your application is more likely to go through with fewer hitches.

You need a permit any time you alter the existing system. This includes …

➤ Adding additional receptacles or fixtures.

➤ Adding new circuits.

➤ Installing a new service or a subpanel.

You usually do not need permits for repairs or updates that do not extend the existing system. This includes …

➤ Replacing existing lights with new fixtures.

➤ Replacing broken switches or receptacles.

➤ Replacing defective circuit breakers.

Always consult your local code to confirm whether you need to take out a permit before you do your work.

Keeping the Inspector in Mind

An electrician has some built-in advantages when dealing with an electrical inspector. He or she can speak knowledgeably about code issues and can justify the manner in which the work is being carried out if there are any questions or objections. It's assumed that an electrician will be basically competent. As a do-it-yourselfer, your work should be neater, cleaner, and on the conservative side. You don't want to be pushing the electrical-code creativity envelope.

The Code Calls the Shots

This bears repeating: Your local code lays out the rules you must follow when doing electrical work. It might seem very logical to you, for example, to add a bathroom receptacle by extending your underused bedroom circuit. After all, there's a bedroom outlet right on the other side of the bathroom wall. What could be simpler than running a few wires and calling it good? Simpler or not, the code states the receptacles in bathrooms must be dedicated to the bathroom, be GFCI protected, and have a minimum ampacity of 15 amps. Safety, not simplicity, is the main concern of the code.

Electrical Elaboration

Remember, the inspector has the final word, whether you're a do-it-yourselfer or a been-in-the-business-forever electrician. This doesn't mean an inspector can't make a wrong call, but the call must be changed in your favor, or you must resolve the problem as the inspector sees it. Some inspectors have reputations for simply being disputatious; others are more reasonable. In the worst cases, disputes over work are taken to an inspector's supervisor.

Insurance Issues

If safety and legality aren't compelling enough reasons for you to follow your local electrical code and have your work inspected, a chat with your insurance agent might be more convincing. Any damage to your house as the result of faulty, uninspected electrical work (no matter whether it's done by you or an electrician) probably will not be covered by your insurance company (read your policy carefully). This can range from something relatively simple such as cleaning up after smoke damage to losing your entire home in a fire.

Let this sink in for a moment. You could lose a $200,000 house if a kitchen circuit isn't installed properly, becomes overloaded, and starts a fire in your walls—all due to the lack of a permit and inspection. Why take the chance? Some electricians might tell you not to bother with a permit for some jobs, but you have to wonder why a contractor would put his business reputation or even license on the line like that. Gambling is more fun in Las Vegas. At least you can get a cheap steak dinner and 24-hour lounge acts out of it.

Tools of the Trade

Neanderthals first used stone tools around 70,000 B.C.E., and life has never been the same since. What started out with a guy named Og making a few simple carving and cutting implements has grown into a multibillion-dollar manufacturing behemoth. From fine Japanese woodworking saws to portable cement mixers, there isn't a single tool we cannot buy or rent. If you walk into any Home Depot or other large home-improvement chain store, you'll find an absolute cornucopia of both hand and power tools.

Every remodeler needs some of both. The notion that the builders of yesteryear felt greater personal accomplishment because they did everything by hand is an absolute myth. Workers in the trades grabbed just about every labor-saving power tool they could as they were invented.

Bright Idea

Yard and estate sales can be good sources for used tools. Look for tools in the best condition, especially hand tools. The cutting edges should be sharp, and the ends of the screwdriver blades should not be worn or rounded. Turn on electrical tools and look inside the vented section. You don't want to see any excessive sparking or smell a "burning" motor.

You don't need the very best tools. Some professionals even shy away from top-of-the-line products because they have more opportunities to damage them on a job site or to lose them, sometimes through theft, as they move around to different locations. On some very large jobs, contractor bids might include the price of new tools. Ironically, a homeowner, who will use tools far less often than a professional, probably would get more use out of expensive, top-quality tools.

Electrical work calls for some specialized tools as well as some multiple-purpose power tools.

Hand Tools

You'll already be familiar with some of these tools; others are limited to the electrical trade. These tools will cut, strip, and twist wires and will secure electrical boxes, light fixtures, switches, and receptacles. You *could* strip away insulation with a pocketknife and cut wire by bending it back and forth until it breaks, but you'll end up with sloppy results and damaged cable. Good cutting and stripping tools prevent wires and insulation from getting nicked and enable you to work with wire in tight areas such as small boxes. The following tables list the basic hand tools for electrical work and some more specialized tools, too.

Basic Hand Tools for Electrical Work

Tool	Purpose
Claw hammer	Securing boxes to studs and joist
Long-nose pliers	Bending wires
Lineman's pliers	Pulling wires and cutting
Diagonal pliers	Cutting in tight spaces
Slotted screwdriver	Securing switches and receptacles
Phillips screwdriver	Securing switches and receptacles
25-inch measuring tape	Setting box heights and so on
Keyhole saw	Cutting through walls and ceilings
Hacksaw	Cutting flexible, armored cable
Wire stripper	Stripping wire insulation
Cable stripper	Stripping cable insulation
Flashlight	Working in dark spaces
Ladders	Accessing overhead work
Voltage tester	Testing for current
Receptacle analyzer	Testing for electrical faults
Continuity tester	Testing for interruptions in the path of a current

Specialized Hand Tools for Electrical Work

Tool	Purpose
Fish tape	Pulling wires through enclosed areas
Conduit bender	Bending metal conduit around corners
Level	Ensuring that equipment is installed straight and true
Masonry chisels	Working on exterior installations
Adjustable wrench	Tightening rigid conduit connectors

Your task will determine the tools you'll need.

Power Tools

During the dawn of electrification at the end of the nineteenth century, electric tools had yet to be invented. Knob-and-tube wiring passed through wall plates (the horizontal 2×4s at the bottom of a wall) but ran along the surface of the wall studs, so electricians had little drilling to do. These days, it all passes through studs, plates, and joist, and no one in their right mind would hand-drill the necessary holes. The electric drill is probably the most ubiquitous power tool around.

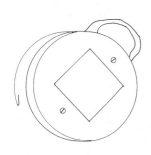

fish tape

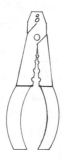

multi-purpose wire stripper

continuity tester

screwdrivers

pliers

These are some of the hand tools you'll need for electrical work.

(Photos courtesy of Craftsman)

Positively Shocking

Cutting tools must be kept sharp; otherwise they can slip and injure you or damage the work. Screwdrivers also can slip and should be examined for dulled or rounded blade ends. Pliers can turn smooth and can lose their gripping capability. Any tool that cannot be repaired should be replaced.

Drills are manufactured according to chuck size. The chuck holds the drill bit or another attachment such as a grinding wheel or a buffer pad. The larger the chuck, the bigger the drill motor (because more power is required to drive larger drill bits and attachments). A manufacturer's usual line of drills includes $\frac{1}{4}$-inch, $\frac{3}{8}$-inch, and $\frac{1}{2}$-inch models.

Drills come in both corded and cordless models. A cordless model runs on a rechargeable battery. Cordless tools are really convenient. The drawback, of course, is battery life. The tougher the task, such as drilling through wood joist, the more demand on the battery and the shorter its work life before needing a charge. A high-end model such as a Makita $\frac{1}{2}$-inch, 18-volt cordless will hold up longer under more demanding drilling, but it isn't cheap (around $245). For repetitive, serious drilling, a corded model often is the best choice.

A cordless drill usually costs more than a corded model of equal size. Sales and close-outs can narrow the gap. Power is the key here, so stick to at least a 12-volt model. Also compare charging times for the batteries (the shorter the better).

A decent ½-inch drill will get you through most electrical drilling chores and just about any other job around the house. It will last for years and years doing occasional residential work without breaking your remodeling budget. Prices for an acceptable ½-inch electric drill start at around $70 and up. You'll find a few other power tools and accessories to be useful as well including ...

➤ A reciprocating saw for sawing large holes in walls and ceilings.

➤ A circular saw for framing work.

➤ 12/2 extension cords.

Extension cords are manufactured according to wire gauge, just like electrical cable. A 12/2 cord is made from 12-gauge wire and contains hot, neutral, and grounding conductors.

Bright Idea

Buy a second battery along with your cordless tool if you expect to be using it extensively. This way, you always have a battery charged and ready to go while the other battery is being used. Professionals almost always have two batteries on hand and rotate them.

Corded and cordless drills.

(Images courtesy of Makita USA, Inc.)

Reciprocating saws.

(Photos courtesy of Makita USA, Inc.)

Care and Feeding of Power Tools

Power tools are great timesavers and are more fun to use than hand tools. These tools won't be fun for long, however, if they're misused and abused. Be sure to avoid the following:

➤ Lifting the tool by pulling on the power cord instead of the handle or body of the tool

➤ Dropping the tool, especially from the second floor to the first

➤ Applying too much pressure while using the tool and ignoring warning signs such as the blade or drill bit slowing down and straining or the motor giving off a burning smell

➤ Ignoring damaged cords

➤ Leaving the tool out in the rain

Tools don't ask for much. They're like huskies and dogsleds. If you treat huskies well and keep them fed, they'll pull your sled until they drop. A power tool will keep going and going if you take care of it. I've run across homeowners with 40-year-old electric drills that still run like the day they came out of the box.

Taking care of your tools also will protect you. A frayed cord can lead to an electrical short, which is not good for your health. A dull blade or drill bit can cause the tool to slip and cut you instead of the wood you're aiming at.

Electrical Elaboration

It's best to recharge your drill batteries as soon as the drill begins to slow down. It used to be recommended that the battery be completely run down to get a full charge, but this no longer is the case and in fact can ruin the batteries.

Bits

Drill bits come in every shape and size for all types of jobs, from drilling through masonry to fine craftwork. The most common bits most of us have seen are twist bits that are sold both individually and in sets based on gradation. Twist bits are fine for small holes, but they're not much use in electrical work except for running small, low-voltage wires.

Electricians run more than one cable through a hole whenever possible, and larger holes (one inch in diameter) are drilled with another type of drill bit. The following are the most common bits for drilling larger holes:

➤ A spade bit

➤ An auger bit

➤ A power bore bit

When running a single cable through a wall stud or joist, a ⅝-inch hole usually is drilled.

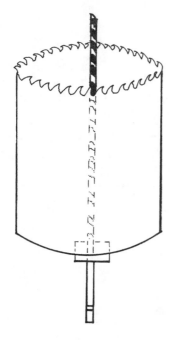

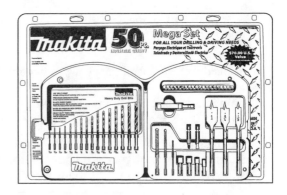

A drill bit and hole saw for everything.

(Images courtesy of Makita USA, Inc.)

Rent or Buy?

Tools might come with some of the same specifications, but one $\frac{1}{2}$-inch drill isn't necessarily the same as the next. One of the main differences is the size of the motor. Professional, heavy-duty models have large motors and can reduce your drilling time through wood and masonry. Hand tools have their differences as well, usually in the quality of the metal components and the sharpness of the cutting edges. Price is a good determinant here, and you really do get what you pay for. That's why those 99¢ screwdrivers lose their square edge quickly.

Bright Idea

Be sure to check online prices for power tools. You'll at least get some idea of whether a local price is reasonable. Tool prices generally are pretty competitive, but check a few different sources before making your purchase.

If your framework is exposed and you're ready to do a lot of drilling at once, you might be better off renting a drill with a large motor. An electrician will use a heavy-duty drill regularly and can justify the expense of owning one. The larger the motor, the weightier the drill, so these tools aren't appropriate for all drilling jobs unless you're on steroids. As a cost comparison, a Makita 7.5 amp, $1/2$-inch angle drill (a large right-angle drill) runs about $290. Local rental prices vary, but I'd be surprised if it cost more than $20 or so a day to rent one of these. A right-angle drill is convenient for drilling in tight spaces.

Before you rent a tool, handle it and get a feel for it. A heavy tool can be uncomfortable to hold for an extended period of time. You might be better off with a smaller drill that you can handle more safely. When I had nothing else available, I drilled through old, hardened floor joist with a $3/8$-inch drill without the sharpest of drill bits, and I still got the job done (not that I recommend this approach—it took time).

UL-Approved Parts for You

It would be unusual to run across an electrical component that isn't UL (Underwriters Laboratories) approved, but always check for this tag or stamp of approval on anything you buy, whether it's a flashlight, power tool, or electrical device. A UL listing is your assurance that the product has been tested for safety. Receptacles, light switches, light fixtures, and appliances all should have UL approval.

Keep in mind that UL approval doesn't imply longevity or ease of installation. A cheaper, lower-end product will never be the equivalent of a more expensive product.

Home-Improvement Stores vs. Electrical Wholesalers

Electricians usually do their shopping with suppliers whose stock and trade are electrical components. Large commercial companies always shop at these establishments. They don't share their space with paint, kitchen cabinets, or coupon specials. They also don't operate quite the same as a retail establishment. This means …

➤ Don't expect them to explain how to do your wiring or use the tools.

➤ Very little will be on display, so you'll need to have a clear idea of what you want when you go in.

➤ If a store is wholesale only, it might sell only to contractors.

Electrical Elaboration

I was in one Seattle electrical wholesaler's store some years ago when a lot of electricians began streaming in on their lunch hour. A homeowner was next in line, and he came in to buy an entire electrical service. He wanted to know what he needed, explaining that he had called earlier and was told they would "fix him right up." The clerk was not amused as he peppered the customer with questions about the service size he required, the type of meter, whether the cable was coming in overhead or buried, and so on. Answers were not forthcoming, and the line of electricians, who knew what they wanted, just kept growing ....

The advantage of shopping at a wholesale supplier used to be the range of supplies and devices available, but in recent years, the large homeowner-oriented building-supply stores such as Home Base and Home Depot have narrowed this gap in the residential categories. A trip to our local Home Base found Square D, Cutler-Hammer, and West-inghouse service panels available as well as a huge variety of boxes, conduit, cable, connectors—you name it. These stores are geared toward do-it-yourselfers who need to see the components and fixtures instead of trying to order them blind at a wholesaler. They also order merchandise by the trainload and usually have competitive prices.

The Least You Need to Know

➤ As a do-it-yourselfer, you want to handle your electrical work at least as neatly and as safely as a professional electrician.

➤ Careful planning always saves cleanup and patch-up time.

➤ The very best tools might be out of your budget range, but the cheapest will cost you more. Find a happy medium when you go tool shopping.

➤ It's easier to take care of your tools than to replace them.

Chapter 9

Extension Cords and Multiple Strips

In This Chapter

➤ Differences among extension cords

➤ Choosing the right cord

➤ Avoiding overloads and problems

➤ Protection from surge suppressors

During large construction and remodeling jobs, it's common to see extension cords snaking throughout the work site. In fact, the condition of individual extension cords is a common safety violation on these sites. Cracked insulation, loose plugs, and broken or missing grounding pins are the most common problems.

A safety inspector isn't going to ring your doorbell and check out your use of extension cords or multiple-outlet strips. Despite their common use, many people don't understand extension cord protocol. Something as simple as the placement of an extension-cord can make it a safety hazard. Some manufacturers don't help much, either. Thousands of potentially defective cords have been recalled in recent years.

Surge suppressors offer more than just a few extra outlets at the end of an electrical cord. They can protect the valuable electronic equipment (such as computers, sound systems, and fax machines) that we increasingly can't live without from voltage jumps in our electrical systems. Think of a surge suppressor as the sacrificial lamb that takes the hit from a voltage spike so your computer can live to frustrate you another day. Unfortunately, a suppressor cannot work without a grounding conductor, so if you have an old two-wire system, you're out of luck.

Every manufacturer of this protective equipment seems to believe its suppressors could soak up a bolt of lightning thrown by Thor himself. A little information and a few figures will help you choose the best suppressor for your home needs.

Extension-Cord Protocol

Think of an extension cord as a portable, impermanent form of wiring. It's subject to the same laws and limitations as any other electrical conductor, which means it can be overloaded, it can short out, and its insulation can melt. On top of that, they're easy to trip over when they're left lying around a work site.

Extension cords are handy and necessary, but they need to be used carefully and inspected before each job. According to the Consumer Product Safety Commission …

➤ More than 3,000 residential fires each year are attributed to extension cords. Most of the problems are the result of short circuits, excessive loads, and misuse of the cords.

➤ Hospital emergency rooms treat more than 2,000 injuries each year associated with extension cords. These injuries include fractures, lacerations, and sprains from tripping over the cords. About half of the injuries to children are caused by electrical burns to the mouth.

➤ Tens of millions of dollars in fire damage occur yearly as the result of misuse of extension cords.

You might never look at an extension cord in the same way again!

Bright Idea

Unplug and put cords away after each use. This happens on most commercial job sites as part of the daily cleanup before everyone goes home. It's a good practice for a do-it-yourselfer as well.

What the NEC Says

Extension cords aren't permanent wiring, so the National Electrical Code doesn't apply to them per se, but it does lay out some guidelines. The NEC would prefer that you use a close-by receptacle, but that's a little unrealistic if you're running an electric lawnmower, for example. The code recognizes that extension cords are meant only for temporary use for portable loads that aren't fixed to one specific location.

What about a table lamp and clock radio by your bedside connected to a small-gauge extension cord because the only receptacle on the wall is beyond the length of their appliance cords? The NEC would like you to install a new receptacle, but sometimes this isn't practical. If you're renting, you would have to convince your landlord/landlady to accept this additional expense. For such a small load, the use of an extension cord on a regular basis isn't a big deal. Problems occur, however, when extension cords are used to run large loads on a more or less permanent basis. Small extension cords become a bigger problem when they are installed under a rug or are in any way covered over so they retain heat.

One Size Doesn't Fit All

Extension cords are measured by their wire gauge size just like the wires running inside your walls. Their ampacity rating uses the same *American Wire Gauge (AWG)* standards: the smaller the number, the thicker the wire (which means it can carry a larger current because it offers less resistance). This is especially important with longer cords because a current loses some voltage as it travels over a conductor, and this can affect the performance of a device (such as a power tool) on the other end. When the voltage drops, any electrical equipment on the cord will pull more current to compensate for the lost voltage. This generates more heat, which causes damage to the tool. The longer the conductor, the greater the voltage drop. Contractors usually use a 12/2 extension cord to run their tools, and you should, too.

Typically, extension cords are available in 10, 12, 14, 16, and 18 gauges. An 18-gauge cord is the size normally used for very small loads such as lamps or clock radios.

The following table shows typical extension-cord lengths and gauge combinations.

Positively Shocking

You don't want to overload an extension cord by plugging in an electrical load too large for its ampacity. One sure sign is a cord that's warm or hot to the touch. If this happens, turn the devices off or use a larger cord. And *never* use an extension cord to run a portable heater!

Ask an Electrician

AWG stands for **American Wire Gauge.** You can always use a larger-gauge wire than a load calls for, but you cannot safely use a smaller gauge. You can't go wrong with a 12-gauge extension cord for your remodeling work.

Extension Cord Gauges (AWG), Lengths, and Amp Ratings

Amp Rating	Cord Length		
	50 Foot	100 Foot	200 Foot
8 to 10 amps	18 AWG	14 AWG	12 AWG
10 to 12 amps	16	14	10
12 to 14 amps	16	12	10
14 to 16 amps	16	12	10
16 to 18 amps	14	12	8

When Cords Go Bad

In the 1960s, the first recalls of defective or assumed-defective automobiles began, and we've been recalling consumer products ever since. Surprisingly, there have been a number of extension-cord recalls, although they don't get quite the same publicity as, say, recalling the family minivan because the wheels have a tendency to fall off. The following is a list of recent recalls, courtesy of the Consumer Product Safety Commission:

➤ Approximately 230,000 extension cords manufactured in China and distributed by a Texas firm were recalled due to undersized wires and improper plugs, according to a September 20, 1994, announcement.

➤ In a May 29, 1997, press release, the General Cable Corp. announced the recall of 2,700 outdoor extension cords sold under the Carol and Ace brand names due to an exposed wire near the receptacle. No injuries had been reported from the use of these cords.

➤ A Miami, Florida, firm recalled almost 6,600 extension cords and power strips in 1998 due to undersized wires that could not carry the advertised load, improperly polarized plugs, and no overcurrent protection in the surge protectors.

➤ A February 24, 1999, press release warned consumers about two *million* faulty extension cords, power strips, and surge protectors involved in 25 recalls since 1994. An ongoing investigation started in 1997 found that most of the faulty cords were made in China, were sold at discount stores, and in some cases, had counterfeit UL certification labels.

This doesn't exactly inspire confidence, but it's useful because it provides the motivation to inspect and check your extension cords on a regular basis. On large construction sites, monthly testing of extension cords for grounding is mandatory, as is recording the test results. Cords that pass inspection are marked with a piece of colored tape. (The color changes monthly.)

Know the Rules

Extension cords come with warnings and usage guidelines just like every other consumer product. Some of the best advice comes from fire departments and the Consumer Product Safety Council, both of which have experience with the injuries and destruction caused by misuse of extension cords and power strips. Here's a list of extension cord do's and don'ts:

➤ Use extension cords for temporary use only.

➤ Unplug extension cords when they're not in use.

➤ Only use cords having gauges that are properly matched to the load and the current to be drawn.

➤ Only use cords outside that are specifically marked for this type of use.

➤ Only use polarized receptacles with polarized cords.

➤ Never remove the grounding prong from the plug of an extension cord to fit it into an ungrounded receptacle.

➤ Regularly inspect your cords for damage.

➤ Never splice a damaged extension cord or one cord to another.

➤ Do not run cords across or through wet areas or puddles.

➤ Hang cords high off the floor to avoid tripping hazards on work sites. Don't allow cords to hang from counter- or tabletops where children can pull on them.

➤ Cover any unused sections of the cord's outlet end with safety caps to keep children from inserting objects.

➤ Replace damaged or worn cords.

➤ Always stretch out the cord, and never cover it with rugs, carpets, clothing, or heavy objects. Cords can build up heat if they are used when coiled or looped.

115

➤ Extension cords are temporary wiring. Don't attach them to walls or woodwork with staples or nails that can damage the cord and present a fire hazard.

➤ Don't plug extension cords together. Instead, use a single cord long enough to do the job on its own.

➤ Buy cords that have been tested by an approved testing lab such as Underwriters Laboratory (UL) or Electrical Testing Laboratories (ETL).

Homemade Cords

Once in a while, I run across a home in which a contractor used extension cords made from nonmetalic cable with a plug spliced on to one end and a pair of receptacles in a box on the other. You would never see this on a large commercial job because it would be considered a safety hazard. The more creative guys attach the receptacles to a wood stand with a plywood base so a work light can be mounted. I don't see any clear advantage to messing around with made-on-the-job cords like this when an approved 100-foot, 12/2 extension cord can be purchased for a modest amount of money and will last for years.

Multiple-Outlet Devices

If you've got a computer and its peripherals (a printer, scanner, and ZIP drive), you probably have a power strip of some kind unless you specifically wire a room of your house as you would an office and give yourself plenty of receptacles. Power strips usually are rectangular-shaped with four or more individual outlets and a built-in circuit breaker.

Most power strips in hardware and discount stores are an all-purpose type and are not appropriate for computer use. Leviton, for example, makes computer-grade strips that, according to the catalog, feature *"EMI/RFI noise attenuation for microprocessor-driven electronic equipment."* They also provide surge suppression.

A Leviton power strip.

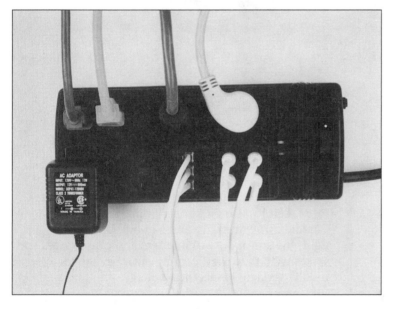

A Leviton suppression strip.

You can add receptacles by using outlet box lampholders. These typically are porcelain light fixtures with pull chains to control the lights (instead of a switch). Some of these lampholders, often found in unfinished spaces in a house, come with built-in outlets. These *really* are meant as a temporary power source, not for running multiple power tools. Remember, as a lighting circuit, it's most likely running on only 15 amps. Note that in a crawl space, the receptacle on a porcelain lampholder must be GFCI-protected.

A Leviton lampholder.

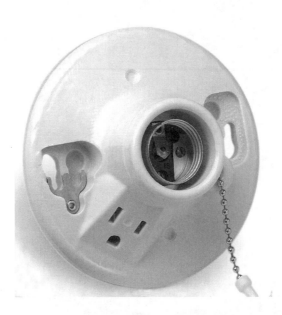

Bright Idea

Consider using a plug-in GFCI outlet that can protect old two-wire systems. These units simply plug into existing receptacles for standard ground-fault protection and can easily be removed and installed in another receptacle as needed.

Surge-Suppression Devices

Electricity is like human relationships: It has its peaks, its low points, and a lot of time in between when it just muddles through without causing much excitement. Left to its own devices, it probably would muddle through day in and day out, but we (and a lightning storm or two) interfere and cause surges and spikes. What are they? Simply put, they are increases, usually sudden, in electrical voltage.

Surges differ from spikes in part by how they occur. Surges can result from ...

➤ The energy demand when a large appliance is first turned on.

➤ Routine maintenance and switching by your utility company.

➤ The rush of current to your house after power that was cut off is turned on again.

Spikes, on the other hand, most often are caused by lightning or by cars running into power poles. Spikes can send as much as 6,000 volts down your line.

A surge is apparent when you turn on a garbage disposer or a laundry-room appliance. The appliance requires a surge of power to get rolling. This is the same principle behind moving a stationary body: The initial force or power required is greater than the amount needed to sustain movement. The first few pushes you give your kids on a swing require more energy than later ones after momentum has been established.

Electrical Elaboration

Another problem, the opposite of a surge or a spike, is a **sag.** Sags are brief decreases in voltage level, and they usually result from too much demand for power on the electrical system. This sudden loss can cause computer crashes and possible loss or corruption of data. A sag is the same as undervoltage. Brownouts and blackouts are huge undervoltage situations.

Both surges and spikes are fairly short in duration, but they differ in their voltage consequences. A spike often brings thousands of volts with it but only lasts for a few microseconds (millionths of seconds). A surge lasts longer but has much lower voltage as a rule. Your electrical system isn't quick enough to protect your relatively fragile electronic equipment from surges and spikes, but a surge suppressor will do the trick.

What Do They Do?

In addition to making surge-suppressor salespeople happy when they sell them, a surge suppressor protects all the electronic equipment we seem to have around our homes these days, from computers to VCRs. A suppressor, which typically goes between your electronic equipment and a receptacle, detects a voltage increase and prevents it from continuing into the equipment.

Bright Idea

You can't predict when a spike will occur, but a lightning storm rolling in serves as a warning. If you don't need to use your electronic equipment at the time, simply unplug it until the storm is over. If you're leaving town for an extended trip, unplug everything electronic and you won't have to worry about it.

Positively Shocking

A surge suppressor can only go so far in protecting your equipment. It won't stop a lightning strike to your power lines, but a lightning rod might. Another legacy of Ben Franklin, a lightning rod redirects lightning away from your house and into the ground.

What should you consider protecting? Anything with a microchip such as …

➤ Computers.

➤ Televisions, VCRs, and stereos.

➤ Telephones and answering machines.

➤ Microwave ovens.

Why telephones? Phone lines and cable lines run in close proximity to power lines. An electrical surge could travel down the phone or TV line instead of the power line. Not only your phones but also your computer could be damaged if it has a modem and is connected to your phone line.

Computers Aren't Very Tough

We can replace a television or an answering machine quite easily, but it's not so easy to retrieve lost data on a hard drive after a surge or spike hits the old PC. The worst-case scenarios, at least as presented by various surge-suppressor manufacturers, include …

➤ Losing any data in memory.

➤ Possible damage to the file allocation table because the computer would not have been shut down properly.

➤ The stress of regular, unnoticed surges gradually deteriorating your computer's components.

➤ Sags causing a system crash and the possible loss of data.

➤ A strong spike frying your PC.

If you live in an area of frequent storms or windy weather that might cause your power to go out, a surge suppressor should be higher up on your birthday wish list than if storms are infrequent. Nevertheless, given our creeping dependence on computers and stored data, a surge suppressor is a good idea wherever you are. (Sorting through all the competing claims by different manufacturers is another issue altogether.)

Suppressors for Everyone

If you search the Web for surge suppressors or go into a computer store, you'll be surprised at how many manufacturers have the absolutely best product available. They can't even agree on the best criteria to judge suppressors. What's a consumer to do? Simple: Pick the suppressor with the best warranty. Such a warranty will guarantee the following:

> ➤ Repair or replacement of the suppressor and any connected equipment *for life* if the suppressor fails to protect against surges
>
> ➤ Payment for the retrieval of lost data
>
> ➤ A high maximum dollar amount for damages

This kind of guarantee doesn't come as cheaply as a less-inclusive one, but it does make choosing a suppressor a lot easier than wading through the claims and manufacturers' specifications. If you must wade, here are some considerations before you purchase:

Bright Idea

Don't take a chance on losing any of your computer data, even if you have a top-quality surge suppressor. Follow the golden rules of computing by regularly saving your data and making backup copies. No lightning bolt is going to jump out of your wall receptacle and cook a ZIP disk after it's been removed from its drive.

> ➤ The suppressor should have at least a UL 1449-330-volt let-through rating (the lowest amount of voltage the suppressor allows to pass through). This is basically a safety rating. A higher rating, UL Adjunct Endurance Testing, meets tougher government Commercial Item Description (CID) Class, Grade, and Mode specifications.
>
> ➤ Telephone line, fax line, and coaxial cable line protection should be provided.
>
> ➤ It should have a high joule rating. (This measures your suppressor's capability to absorb energy, which is measured in joules.)
>
> ➤ It should have high surge amp ratings.
>
> ➤ The suppressor should have an indicator light to show that the device is working.
>
> ➤ It should provide protection in all three modes (surges between hot, neutral, and ground lines).
>
> ➤ It should have instantaneous response time.
>
> ➤ The unit should shut off power to all of its outlets once the unit has reached its capacity to protect.
>
> ➤ It should offer a broad degree of EMI/RFI noise reduction.

Competing claims among surge-suppressor manufacturers start sounding like taunts between opposing cliques in the schoolyard. It begins to sound like the Macintosh versus PC battle. My advice? I'd still buy the suppressor with the best guarantee for my price range. Any damage to your equipment or data then becomes the manufacturer's problem. (Read the guarantee carefully.)

Speaking About Computers

A couple other tech toys you might not have known you desperately needed are line conditioners and an uninterrupted power supply (UPS). A line conditioner adjusts the line voltage to a norm, getting rid of highs and lows. This is a good piece of equipment to have if your home electrical system has a regular case of the sags. A UPS is basically a sophisticated battery pack with various filtering properties that kick in when you have a power outage. The key word here is "battery." You don't want to be running your laser printer or copying machine off of this if you lose power. Use it for your computer, not the peripherals.

More Rules

Surge suppressors come with a few guidelines as well, just like extension cords. These guidelines include …

➤ Don't go beyond the electrical rating of the suppressor.

➤ Surge suppressors are designed for indoor use in dry areas.

➤ Don't plug the suppressor into an extension cord.

➤ Keep children and pets away from the suppressor's power cord.

➤ Suppressors are not designed to be used with aquariums.

Going Whole Hog

Some available systems offer protection starting at your home's meter. A suppressor is installed near the electric meter, and it protects major appliances from surges and lightning strikes. Standard plug-in suppressors are installed inside your home for more sensitive equipment. The Square D company manufactures an inexpensive surge suppressor that mounts directly in the service panel like a two-pole breaker. Talk with your utility company to inquire about these whole-house systems.

The Least You Need to Know

➤ Extension cords are widely used, but they are not always used appropriately.

➤ For most jobs, a heavy-gauge extension cord is a safe choice.

➤ Give your extension cord a quick inspection before, during, and after a job.

➤ Sorting out competing surge-suppressor claims can be confusing. Always look for a solid guarantee against damage and data loss.

Chapter 10

Electing for Electricians

In This Chapter

➤ You and your contractor

➤ Working on the plans

➤ Understanding your contract

➤ Knowing who does what

➤ Choice materials, choice results

Few homeowners do all the necessary electrical work on an old house. Upgrading a service, tying into old circuits, and rewiring existing ceiling lights can be intimidating tasks (of course, that's why you bought this book). Even if you choose to do more limited electrical work and hire the rest out, a good working knowledge of electricity and your home's electrical system will enable you to discuss the job intelligently with your electrician and to compare bids more critically.

When you hire a contractor, you each have your respective responsibilities and expectations. You need to clearly communicate what you want done and the time frame in which it must be completed. The contractor must be equally clear in stating the work as he or she understands it from your plans, the cost for labor and materials, and a reasonable completion date. Any changes by either party must be negotiated.

This might be a new experience for you. You'll find this stranger and perhaps a crew of one or two people wandering around your house in work boots, punching holes in the walls, and shutting your power off from time to time. Who are these people and how do you deal with them? Suddenly you're an employer of sorts, hoping these new employees are going to work out before you write them a check.

You and your contractor should have the same goal: an efficient job done as agreed to in advance with a minimum of disruption. Don't worry, clear communications with a carefully selected electrician—and maybe a box of doughnuts in the morning—will smooth the way for everyone involved.

Hiring It Out

The most meaningful compliment I received about my last book, *The Complete Idiot's Guide to Remodeling Your Home,* came from the production editor. She read the book while in the process of buying her first house and hiring a home inspector to give the place a once-over. He told her he was impressed with how much she knew. Her knowledge allowed her to intelligently scrutinize her potential new home and more clearly understand the inspector's comments and observations.

You'll want to be knowledgeable as well, not only about your electrical system but also about contractors. A contract, whether it's oral or written, is a legally binding agreement. You need to know your rights, the contractor's rights, the bidding process, and payment schedules. There also are intangibles such as your personal reactions to individual electricians bidding the job. If red flags start popping up in front of your eyes, you should start looking for another electrician.

First, however, you have to *find* an electrician.

Bright Idea

In prosperous times, contractors tend to be booked solid weeks and months in advance. Even getting an estimate can be difficult, especially during the summer months. Plan ahead as far as possible.

Contracting for a Contractor

If you skim the Yellow Pages, you'll find lots of listings for electrical contractors, but that's not the best way to choose one for your job. You probably didn't find your physician, dentist, or auto mechanic this way, so why choose an electrician blindly? Do what you did with all the other professionals in your life—get some referrals.

Start with other homeowners. They will be your most obvious resource, particularly if they've done any remodeling. Ask your friends, family, co-workers, even your dentist! There is no guarantee that a contractor will give you the same results in your home, but there's a good chance you'll be satisfied with the results of a referral. Most small contractors survive on referrals and will want yours as well.

A contractor's time is valuable, so don't call a dozen of them to give you a price for adding one circuit to your house. A larger job (such as a service change or a total rewire) is another matter, and three or four bids would not be inappropriate. First, however, there are a few legalities to consider.

License and Bonding Spoken Here

Unless you live buried away in the extreme northeast corner of Montana in an area so remote that no one, not even the IRS or junk mailers, knows it exists, you should expect an electrician to be licensed, bonded, and insured in accordance with local and state laws. These requirements are fairly standard across the country. They protect you and the contractor from each other if problems arise.

A license is simply permission from a governing authority to do a specific business. It shows that a contractor is registered, often with both the city and the state, and has met certain standards. This enables a contractor to hang a shingle out and say, "I'm an electrician." It also means the local government has collected a registration fee and will be collecting taxes from the licensee.

Two requirements usually have to be met before a contractor's license is issued:

➤ The individual must be bonded.

➤ The business must be insured.

The Name's Bond, Surety Bond

A contractor's bond (*surety bond*) is required in many states before a contractor will be issued a license to operate. The bond helps guarantee that a contractor will perform according to the terms of a contract. I suppose it's not much different in principle from a jail bond, which is an attempt to guarantee a defendant's appearance in court, but with a more wholesome connotation.

A bond is registered with a governing authority in one of two ways:

➤ The contractor can establish a special account with a cash deposit equivalent to the amount of the bond.

➤ A bonding company can be engaged for a fee.

Positively Shocking

Make sure your electrician's license, bond, and insurance are current. All three should be renewed on a yearly basis. If you have any questions at all, call your city or state department of licensing and do a credentials check. You don't want any problems from hiring an unlicensed individual.

The amount of the bond varies from state to state. In Washington, for example, the bonding rates are relatively low. A general contractor only has to post a $6,000 bond, and a specialty contractor or subcontractor (electricians, plumbers, painters, and so on) must post only a $4,000 bond. If you are not satisfied with a contractor's work, you can put in a claim against the bond, although you're limited to its dollar amount. This isn't much consolation if all you can collect is a fraction of the value of the work, and you must pursue additional financial relief through the courts or arbitration.

Any claim against a contractor must be legitimate. You have to prove that the work was not done to the specifications agreed to in your contract. Just as a bond gives you some leverage in the event of faulty work, a *lien* (sounds like "lean," appropriately enough) gives a contractor some protection against a customer's spurious claims. Sometimes called a *mechanic's lien*, this handy piece of legal work enables a contractor to file a claim against your home until your debt is paid. This doesn't mean your contractor is going to take up residence in your spare bedroom if you don't pay, but the lien must be satisfied before the property can be sold. In some cases, a forced sale of the property can occur.

Insurance Is a Must

I once did an Internet search for insurance jokes, and I couldn't find any. There were plenty of jokes about lawyers, doctors, and even some about accountants, but insurance seems to have escaped the comical wrath of joke writers. This is the bottom line: You want your contractor to be fully insured.

Proof of insurance usually is a requirement for a contractor to obtain a license. Insurance protects you if there's an accident or damage during the course of the work. In addition to a general liability policy, contractors must cover their employees with government-mandated policies such as workers' compensation.

Three in One

A legitimate contractor will be licensed, bonded, and insured. Without all three of these qualifications, you're putting yourself and your home at risk. If a cash-only, unlicensed, we-don't-need-no-stinkin'-contract electrician works on your house and falls off a ladder, shorts out an appliance, or incorrectly wires a circuit that causes a fire, you might never receive compensation for damages. When a licensed electrician causes a problem, you have some legal assurances the problem can eventually be paid for.

Plans and Specifications—Always!

You can't expect someone to bid on a job if you don't specify exactly what you want done. It's not enough to say, "Just add some receptacles and lights wherever you think we need them." You have to specify where you want them, the types of fixtures you want, and even the styles of light bulbs. You don't need detailed plans and specifications for everything. Adding a clothes-dryer circuit, for example, is pretty straightforward once you've designated where the laundry will be located.

Details increase as the scope of the job increases. Installing a new service panel might mean a different location than an existing box. (This obviously is true when an old fuse box located off a back porch is replaced.) A complete update of your existing system, including running all new wire, would have to be detailed, especially when it comes to fixtures and their locations. The following list outlines a very basic plan:

Sample House Plan

Main service: 200-amp Square D QO service panel

Location: NE corner of basement

Existing fuse box will serve as a junction box for any existing circuits to be retained. The door will be screwed shut.

New circuits to be added: Washing machine, dryer

Kitchen: Add two 20-amp small-appliance circuits with GFCIs, white Leviton receptacles, and cover plates. Install nine recessed cans (white trim) with dimmer switch (white) and two 18-inch fluorescent fixtures over counters. Run outlet for range and separate circuits for microwave, refrigerator, and disposer.

Lighting: Add sufficient 15-amp circuits to bring bedrooms, living and dining room, and hallway up to code for receptacles (six-foot rule).

Office: Run dedicated 15-amp computer circuit.

Master bathroom: Run GFCI. Install six-light fixture over mirror and recessed can over toilet (white trim). Install Nutone QT-200 fan.

First-floor bath: Run GFCI. Install four-light fixture over mirror. Install Nutone QT-100 fan.

Living room: Install four wall sconces and one recessed can over fireplace.

Dining: Use existing chandelier. Check wiring for safety.

Bedroom hallway: Use existing fixture and check wiring.

Use existing bedroom ceiling lights and check wiring.

Basement: Run 20-amp circuit for workshop. Install four-foot fluorescent fixtures. Install six ceramic light fixtures in basement ceiling, locations to be marked.

Garage: Run GFCI and one light over each car bay. Run wiring for two garage-door openers (to be installed by others).

Front porch: Install new porch light (Nautilus style). Install GFCI for outdoor use.

Rear porch: Use existing light and check wiring. Install GFCI for outdoor use.

Contractor will supply all labor and materials and will remove any refuse from job site. Job will be kept broom-clean daily. Billing will be done in two installments with a 10-percent down payment to be applied toward materials.

Bright Idea

Large home-improvement centers and specialty retailers are great places to visit to see current types of fixtures and lighting systems. Visit a number of establishments before making your final decisions. Check out disposers, fans, and other small-electrical items you might be installing as well.

Who Draws Them Up?

You, your designer, or your electrician will draw up or sketch any plans for the electrical work. Written descriptions ("locate panel in NE corner of basement") usually are adequate for most residential jobs. Specific light locations, however, should be noted on a sketch or plan of the room. It's not a bad idea to put some kind of marker on the wall, such as blue masking tape, to confirm the location. An architect's or designer's plans for a general remodel should note any electrical requirements.

Allowing Substitutions

As remodeling bids come in and budgets get stretched, your imported marble countertop might suddenly become plastic laminate and your oak floor might become vinyl. The same is true with electrical work. Lights, appliances, and garage-door openers are available in a range of models and prices. Sometimes your electrician can come up with an equivalent-model fixture at a lower price with no appreciable difference in quality or appearance. Your bids and specifications should allow for such substitutions.

Comparing Bids

A clear set of plans and specifications enables all the bidding electricians to play by the same set of ground rules. It also helps you fairly compare their prices. You'll find, as you put a job out to bid, that each electrician has a slightly different take on how to do the work and what materials to use. Keep these suggestions in mind as you scrutinize the bids so you can adjust for specific differences in cost.

Let's say you want a standard, switch-controlled light to be installed outside your garage. One of your bidders might suggest that you put in a motion detector instead, which will automatically turn on the light when it detects someone moving nearby. Another bidder might suggest that you install a larger ventilation fan in your kitchen. The service panel is the big item. If you specify one brand and an electrician recommends another, find out why and compare the differences in cost by calling an electrical wholesaler.

About Those Contracts

Some contractors—and homeowners—want a written contract for everything. This is unnecessary for small jobs, but there's no harm in writing up a short letter of intent. You could say, for example, "Contractor will supply all labor and materials for one new bathroom circuit with GFCI receptacle for the sum of _____ dollars plus applicable tax. Homeowner will take care of any wall repair or patching." For that matter, your contractor might supply a contract form for small jobs with a written description of the work and ask for your signature to confirm your acceptance.

Larger jobs usually require a written contract. If your electrician is hesitant to provide one or to sign yours, find someone else to do the work. No legitimate contractor will shy away from a valid contract.

Write It Down

A contract should include everything you want done. Don't assume that your electrician can read your mind and will install cream-colored receptacles when white is more common. If you have any questions, ask before you sign.

Electrical Elaboration

In addition to describing the job and the materials to be installed, a contract also should state the terms of payment. Some contractors will want a percentage in advance (depending on the size of the job). Some states limit this percentage. You also might be asked to pay to order materials in advance. If the demand for prepayment seems unreasonable (it's easy enough to find out the cost of the materials by calling a wholesaler), find another electrician. You'll know after you get a couple of competing bids if someone is way out of line or not.

Change Orders

A *change order* is a modification to a contract. It can be initiated by either you or your contractor, but it must be agreed to by both. You might decide to add more lights, for example, or a different type of fixture. Your electrician might run across unforeseen problems such as an existing circuit that must be replaced (when you assumed it could still be used). A change order usually means an increase in the price of the job, but this is not always the case. You might decide to eliminate some fixtures or to go for less-expensive ones, thus lowering your overall cost.

The best change order, ideally, is *no* change order. Change orders can delay a job and might cause your electrician to have to undo work completed under your original specifications to accommodate the requested change. No plan is perfect. Remodeling is a fluid experience. As it progresses, you might see things you did not see during the planning stages. A skylight in the bedroom might become more desirable than the track lighting that just went in this morning. Don't laugh, I had a client with more money than sense who did just that. Out came the new drywall and lights; in went new skylights and windows into newly finished rooms. At least he kept the carpenters employed and happy.

A Deal's a Deal

After you've agreed to the job and have signed on the dotted line, you have to hold up your end of things, too. This means ...

➤ Clearing furniture and household items out of the way so your electrician can work.

➤ Keeping your children at a safe distance from the work activity.

➤ Controlling your pets.

➤ Providing access to your house, either with a key or by being home at the start of the workday.

➤ Understanding that your contractor and any crew will need access to a bathroom and somewhere to take their breaks.

➤ Paying your bill in a timely manner. Small contractors are especially dependent on regular cash flow, and you shouldn't unnecessarily delay payment.

Being a good customer is just as important as being a good contractor—all good contractors have stories about customers from hell.

Positively Shocking

Your electrician can rightfully put a clause in your contract that he or she will not assume any responsibility for damage to household items left in the way or not adequately protected on the work site. This includes anything hanging on a wall that could loosen and fall from hammering, sawing, or drilling through the wall.

Cleanup and Wall-Repair Woes

In an existing house, any extensive rewiring will require opening up some walls and ceilings by cutting into the drywall or plaster. Electricians have two conflicting issues here: One voice—yours—says keep the holes small; the other voice—the electrician's—says a larger hole makes the job easier and faster. Guess which one wins out? I'm not against electricians, and no, it doesn't mean they're going to knock a three-foot-by-three-foot hole in your wall just to pull one wire through it. It does mean, however, that you'll have some wall and ceiling repairs to do after the electrician is finished.

Electrical Elaboration

Major electrical work requires more than one inspection. An initial inspection covers the rough-in work, the running and securing of cable inside the walls. When this passes inspection, it's "good to cover," which means the drywall or plaster can be installed. You cannot cover electrical work until it's been inspected, so don't schedule it until after the inspection.

Drywall and plaster repair costs need to be figured into your electrical budget unless you do the work yourself (see Chapter 15, "Working Around Existing Wiring"). It doesn't stop there, however. Your electrician might have to drill through paneling or wallpaper whose patching is a little more problematic. If a room hasn't been painted in many years, the paint will have faded and won't necessarily match up very well with the can of Colonial blue latex sitting in the garage. Figure this into your planning costs so it's not such a surprise later.

Electricians Hate Plaster Walls

I wouldn't go so far as to suggest that plaster walls are a nemesis of electricians, but they're not exactly fond of them. Some plaster, particularly from the Victorian era, can be very brittle and will fall apart easily when being cut through. All it takes is a piece of wood lath vibrating roughly and a four-inch hole becomes a 12-inch crack. Worse yet is metal lath, which can be a mess to cut through. Don't be surprised if a bidding electrician figures in extra cost if you have plaster walls.

Fire Blocks

A fire block is a horizontal 2×4 nailed between two wall studs to slow the spread of fire through the wall. Electricians find them the hard way when they drop a fish tape or chain (used for pulling electrical cable through finished walls), only to find them blocked part of the way down. This requires cutting into the wall near the block and drilling through it to get the cable through. In addition to slowing down the job, it means more wall repairs for you to do later.

The Least You Need to Know

➤ Other homeowners will be your best source of referrals when you're looking for an electrician.

➤ Only hire a licensed, bonded, and insured electrician to work on your home.

➤ Clearly written job specifications help both you and the electrician come up with an accurate, fair price for the job.

➤ Only sign contracts that you fully understand; if you have any questions, ask away.

➤ As a customer, understand your responsibilities to your contractor.

➤ Be prepared to do some wall and ceiling patching after major electrical work.

Part 3

Components and Simple Repairs

Enough with the theories, history, and precautions. Now it's time to do something with your new tools and newfound knowledge. Older homes always have small electrical jobs to do—from repairing a lamp cord to replacing a switch. These are good jobs to start with until you're comfortable working around electricity.

The chapters in Part 3 will walk you through basic repairs and troubleshooting. Because these are small jobs, you won't have to live in the dark with all your power shut off in the event that you don't finish on the day you start the work. The small jobs (such as replacing an old receptacle or switch) will take less than an hour (apprehension factor included).

Repair and replacement mean new components: switches, receptacles, wires, and lights. I'll discuss all the different flavors, from a plain-vanilla-white duplex receptacle to a four-way switch with a brass cover plate. Chapter 15 will show you how to get your cable from one point to another without tearing up your house too much in the process. These chapters won't make you a card-carrying journeyman electrician, but they will give you a much better sense of the work involved and your degree of comfort with it.

Switches and Receptacles

We use them every day, but we don't think about them much. This is a good indication of the reliability of switches and receptacles. A bathroom light switch, for example, might be clicked on and off 10 times a day (depending on the size of your family). That's a few thousand clicks each year, and the switch keeps going and going. If only our computers and operating systems were that reliable.

A switch controls the flow of electricity between a source and an end device such as a light fixture. In a standard modern light switch, a metal arm inside the switch connects the two screw terminals to which the black, or hot, wires are connected. In the "Off" position, this arm moves out of the way and cuts off the flow of electricity along the conductors.

Receptacles don't face as much mechanical wear and tear as a switch undergoes. A common house receptacle is called a duplex receptacle because it can accommodate two plugs. The metal connector between the screw terminals is fixed in place, unlike the movable arm in a switch. Each prong of a plug is held in place by two pieces of spring metal to maintain a solid electrical contact.

Both switches and receptacles can wear out, especially the original ones in an old home. This chapter discusses the most common types and some of the less-common ones as well.

There's One for Every Purpose

The world of electrical devices is quite varied. The light switch in your bedroom isn't quite the same as those at the top and bottom of your staircase. The latter most likely are three-way switches (possibly four-way), which control an electrical load (in this case, the light) from more than one location. You might have an emergency switch that controls your oil furnace or a timer switch connected to a bathroom fan. You need to know one from another when you go to replace an existing switch or install a new one.

Receptacles are no different. You're already familiar with a standard duplex outlet (your house is full of them) and a GFCI receptacle. There also are single receptacles that take one plug, receptacles for clothes dryers and electric ranges that carry both 120 and 240 volts, and special hospital-grade receptacles. Older homes might have original unpolarized receptacles or even some old twist-lock-style receptacles. You have to know what you're dealing with before you replace it; otherwise, you could create a hazardous situation.

As always, follow the unwritten rule of electrical work: Buy only UL-approved materials. The world of Internet trading and crashing trade barriers means more nonlisted devices than ever will be available, but stick with the tried and true, even if your code allows the others. You want some assurances that you're buying a safe product.

Electrical Elaboration

A large appliance receptacle carries both 120 and 240 volts because it has to supply two different loads. A clothes dryer needs 240 volts, but the dryer's timer, lights, motor, and buzzer run on 120 volts. An electric range needs 240 volts but not so its clock and lights.

Switches

The most common switch in your house is a single-pole switch with a toggle marked "On" and "Off." It typically is used to control a light fixture or a receptacle. *Single-pole* simply refers to electricity flowing in one direction. In most cases, one black (hot) wire is connected to one terminal screw (or it might be back wired), and a second hot wire is connected to the other terminal screw, proceeding on to the light. In terms of physics, a *pole* is just one of two opposite points on a magnet that manifest the magnetic properties. (Remember, a spinning magnet, called a *dynamo,* at your utility's power plant creates the electric current.) Unlike older switches, modern versions often come with a ground terminal for the green or bare copper ground wire.

Bright Idea

If you expect to be doing more repairs or modifications to your electrical system in the future, pick up a half a dozen or so extra switches and receptacles to have on hand. It's a small investment to make for the convenience of having them when you need them.

A single-pole switch controls the current to its load from one location only. Other switches control the current from two or even three different locations.

Three-Way Switch

Three-way switches come in twos so you can control a light from two locations. Their most common location is at the top and bottom of a staircase or at opposite ends of a large room with more than one entrance. Three-way switches come with three terminal screws: Two are the *traveler* screw terminals; the third, which is darker in color, is the *common* screw terminal. The traveler terminals connect one switch to the other. The cable that runs between the switches has two hot wires: one neutral, and one ground, as shown in the following diagram.

Four-Way Switch

This one is always found between a pair of three-way switches. You'd have to have a *really* long hallway or a large room needing switch controls from three locations. A four-way switch comes with two pairs of color-matched terminal screws that conveniently connect with color-matched wires from the two three-way switches.

A three-way switch, traveler screw terminals, and a common screw terminal.

(Courtesy of Leviton)

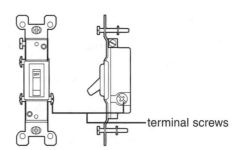

terminal screws

Electrical Elaboration

Switches weren't always the familiar toggle design we use today. A rotary-style switch, which was turned clockwise to open and close a current, was used at the turn of the century. In the 1920s and 1930s, push-button switches were introduced, and they are still in working condition in period homes. Early toggle switches were first manufactured in the 1930s.

Switch/Receptacle Combo

This handy device is half receptacle and half switch. It's a quick way to add a receptacle to a room (after you've calculated the amperage of the new load—remember not to overload your circuits). The receptacle will be at switch height, which typically is four feet from the floor to the top of the receptacle box. The switch and the receptacle can operate independently of each other, or the receptacle can be controlled by the switch, perhaps for a hanging ceiling lamp not directly wired to a circuit that came with a lamp cord and plug.

A naked switch/receptacle.

(Courtesy of Leviton)

Ganging Up

Switches for multiple light fixtures often are *ganged* up in one box. If all the lights are on the same circuit, one feed wire from the panel will supply the power for all the switches and their loads. A separate cable will run to each fixture. Sometimes you'll run across a single gang box with a double switch, but this isn't very common in residential systems.

How many switches can you fit in one box? Leviton offers one switch plate that has space for 10 switches. You'll find these in commercial settings or maybe in an Internet gazzillionaire's new mansion.

Pilot-Light Switch

A pilot-light switch resembles a standard single-pole switch, except it has a built-in bulb (either the toggle is illuminated or the bulb is on the face of the switch) that lights up when the switch is in the "On" position. This usually is installed when the fixture or light is out of sight of the switch (say, a light in a detached garage). The illuminated switch lets you know if someone forgot to turn the lights off.

Dimmers

In addition to their romantic value, dimmer switches enable you to decrease the lighting in the dining room so your kids can't see that you're feeding them Brussels sprouts, a side dish that no human being should ever eat anyway. Dimmer switches come in several styles including those with ...

➤ A toggle control.

➤ A dial control.

➤ A sliding control.

➤ Automatic dimming.

Ask an Electrician

One definition of **gang** is "combining electrical boxes together to form a larger box for additional switches or receptacles." You can buy premade boxes in an appropriate size, or you can combine metal boxes that have removable side walls. Plastic boxes aren't meant to be ganged together and will become damaged if you attempt to do so.

A dimmer reduces the voltage reaching a light fixture, but in doing so, the switch builds up a small amount of heat. Because of this heat and the large size of a dimmer switch compared to other switches, it might not work as a replacement for an existing switch if you have an undersized or crowded box.

Timers and Doorbells

Timers, either manual or automatic, also are types of switches. Manual timers regularly are used with bathroom fans and heat lamps. A frequent residential use of automatic timers is to control a whole-house ventilation system.

Doorbells are switches, too. When you press the button, a low-voltage current flows to the chimes or the buzzer. Thermostats are another low-voltage switch, except these are activated by temperature changes. Elaborate thermostats have separate switches for controlling the furnace fan and for turning heat and air conditioning on and off manually.

Receptacles Galore

Most of the receptacles in your home are the duplex type and have been in common use for the better part of the century. Current versions differ from those used through the 1950s because the newer ones have a grounding hole. Polarized receptacles came into use in the 1920s (see Chapter 4, "If Your Walls Could Talk"). The different-size slots (the longer one always goes with the neutral wire, the shorter with the hot) maintain consistent, directed current flow along the respective hot and neutral wires.

The earliest receptacles were an odd arrangement. The plug-in part of the outlet was actually a screw-in affair, something like a light bulb. The receptacle plate had a small flap that flipped up to reveal a socket into which the plug-in was screwed. (Hey, electrification had to start somewhere.)

Positively Shocking

This bears repeating: Never file down or otherwise alter a polarized plug so it will fit into a non-polarized receptacle. Use your electrical components the way they were designed. If you only have nonpolarized receptacles, use adapters for your polarized plugs.

The following are specialized types of duplex receptacles:

➤ Floor receptacles

➤ Clock receptacles

➤ GFCIs

Floor receptacles are specially designed to withstand foot traffic. They are installed in the middle of large rooms or in other areas far away from a wall receptacle. They often are seen in offices and other commercial settings with large, undivided floor spaces.

A clock receptacle is recessed so that a clock and its cord can be hung flush against a wall. You used to see these more often in kitchens, but you don't see them as often now, especially since the advent of inexpensive battery-powered wall clocks. This type of receptacle is now more common for plugging in microwaves and for picture lights that plug in behind pictures and paintings.

A GFCI can be used for a single location such as a bathroom or a kitchen, or it can offer protection to an entire circuit of receptacles or other loads. This is possible only if the GFCI is the first receptacle on a circuit. From that point on, anything beyond it on the same circuit will have GFCI protection. If it's in the middle of a circuit or in any other position than the first receptacle, it will not offer any protection to any load between it and the service panel.

Don't Forget the Boxes

The NEC code requires that any wires connected to each other or attached to a fixture or device must be enclosed in a box with a cover plate. This means that receptacles, switches, lights, wall heaters, anything that requires electricity will have its wiring housed in a box. A box serves a number of purposes:

➤ It serves as a point of attachment for a device or a fixture. (It has to be screwed to something.)

➤ It keeps wires that could short and then spark away from wood framing, decreasing the possibility of fire.

➤ It protects people from accidental exposure to wires and possible shock.

There are boxes for every purpose: ceiling lights, retrofitting fixtures into existing walls, weatherproof designs for outdoor use, and junctions for wire connections that aren't immediately attached to a device or fixture. As with switches and receptacles, you'll have to choose the right box for the job at hand.

They're Not All the Same

The following is a list of the most common electrical boxes:

➤ Rectangular for switches and receptacles

➤ Square for junctions or two receptacles/switches

➤ Octagonal and round for ceiling fixtures

Bright Idea

Consider buying a package of child-protective caps to stick into any unused receptacles, even if you don't have small children in the house. You never know when you might have overnight guests with toddlers. Some receptacle covers snap tightly in place for something more permanent.

Positively Shocking

Never conceal an electrical box! The NEC states that these boxes must remain visible and accessible, even if it seems unimportant to you.

143

➤ Retrofit types for inserting into existing walls and ceilings

➤ Aluminum and PVC plastic for exterior use

➤ Boxes with extendible bars or braces for attaching between joist

➤ Pancake boxes for limited circumstances when a regular box is too deep for the wall or ceiling space (Most of the plastic ones are not listed as tested by UL, and the metal ones are rated for only one cable, usually 14 gauge.)

➤ Fan-rated boxes—the only boxes you can use to install paddle fans

Boxes can be attached to framing, plaster, and drywall in a variety of ways, as shown in the following figures.

A single gang plastic box with nails.

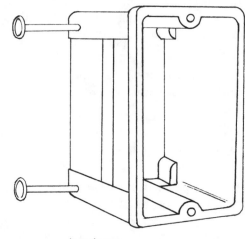

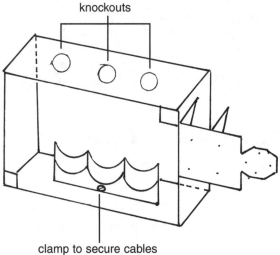

knockouts

A double gang metal box with a bracket.

clamp to secure cables

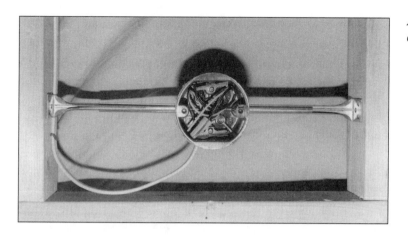

A round box with extendible bars.

Plastic or Steel?

Both plastic and steel boxes are used in residential construction. A box has to withstand a certain amount of construction trauma when it's installed and later when drywall is installed around it. (Drywall hangers are not necessarily kind and gentle people, at least not when they're getting paid by the square foot.) Plastic boxes are lightweight and are easy to install, especially those that come with nails for direct attachment to a wall stud or a floor joist.

Metal boxes are standard in most commercial work. Unlike a plastic box, a metal box is a good conductor of electricity and must be grounded along with the device or fixture. Special fittings are used to connect a metal box to conduit and conductors to the box. These fittings include an array of clamps, clips, and locknuts, most of which you'll never use in the course of residential repairs and remodeling.

A plastic box works well for a single gang or device use, but some electricians find that a larger plastic box's shape distorts during installation or when the drywall is installed. For these reasons, they use tougher boxes, either Bakelite (reinforced phenolic) or metal for two gang installations and metal for three gang. The larger the box, the more difficult it is to keep it level and in line.

Check the Size

Boxes come in different sizes based on the installation need. Rectangular boxes, the most common ones used for single devices, generally are two inches by three inches for residential use. Depth ranges from $1\frac{5}{8}$ inches to $3\frac{19}{32}$ inches. The deeper the box, the more wires it can accommodate and the easier it is to tuck in the wires and install a device and still meet code. You'll really appreciate this when you're dealing with 12-gauge wire.

145

How many cables can your box accommodate? Well, the bigger the box the better, but to be more exact ...

➤ Count the number of intended cables for the box. Each hot and neutral conductor counts as one wire each, and all the grounding conductors together count as a single wire.

➤ Take this total and add one for any cable clamps (if they're the same type of clamp). If you have two different types, you have to count each as a separate number.

➤ Take this new total and add two for each device (switch or receptacle).

➤ If the box contains 14-gauge wire, multiply the total number (of wires, clamps, and devices) by 2 cubic inches. If 12-gauge wire is being used, multiply the total by 2.25 cubic inches.

➤ The result of this multiplication is the minimum allowable volume of wires, clamps, and devices for that box. (The volume of a box usually is stamped on the back of the inside of the box.)

Let's say you have a light switch in a plastic box with two 14-gauge cables coming into it. (One is the line; one is the load.) This gives you two hot conductors, two neutrals, two grounding conductors (these count as one wire in our calculations), and one switch. Therefore ...

Two hot conductors: 2

Two neutrals: 2

One grounding conductor: 1

Device: 2

Total: 7

7 × 2 cubic inches = 14 cubic inch minimum box size

(Note: Most plastic boxes do not have any type of clamp.)

Cover 'Em Up

Every electrical box needs a cover plate. A junction box, which is used solely to house wires and their connections but not devices, needs a blank cover plate. The cover plate keeps probing fingers, especially those of kids, away from the wires and the terminal screws on the device, all of which are fine sources of electrical shock.

Plastic is the material of choice for most cover plates, but metal is used in some commercial work and with metal boxes. Outdoor boxes have plates with foam gaskets to keep moisture out. Outdoor receptacles have additional protection: A section of the cover plate closes over and covers the receptacle when it isn't in use.

Plastic cover plates have been used since the 1920s, but other materials have been used as well.

Brass: New and Old

Many older homes have original brass cover plates, often with a dark bronze tone. These will readily take to a buffing wheel and will come out a fine, shiny brass if that's your preference. New replacement brass plates also are available. Some homeowners and designers install them in kitchens and bathrooms, but such damp locations aren't the best places for brass unless you like polishing them from time to time.

The Artful Flare of Ceramics

If you go to any good-size street fair, at least in a large city, you're likely to run into an artist's booth selling ceramic electrical plates. Some have a theme (such as stars or suns); others are a little more whimsical. Check to see if they're listed by UL; if they're not, decide whether you think it's a problem. These plates are usually ceramic, nonconducting material and can be pricey.

Local gift and design shops might carry these types of plates as well. They usually are purchased for a single room, such as a bathroom or a baby's room, rather than an entire home.

The Least You Need to Know

➤ Every switch and receptacle has a specific purpose; they're not interchangeable.

➤ Despite their long working life, devices occasionally need replacement—a simple job for homeowners.

➤ Every device, fixture, and junction needs an electrical box to meet code requirements.

➤ Cover plates are required for all boxes so that no wires are exposed, especially to small, probing fingers.

Replacing Old Switches and Receptacles

Now that you know about switches and receptacles, it's time to replace any that are broken or to upgrade existing ones. The most common upgrade is swapping a standard toggle switch for a dimmer. Newer, quieter models—that don't have the resounding "click" of old switches—sometimes are installed in older homes that still have their original devices.

The most common reason for replacing a device is wear and tear. The clips in a receptacle that hold a plug tautly or the metal arm in a switch eventually can fatigue and no longer work properly. A simple loss of power to a fixture or an appliance, however, is not necessarily a reason to replace a device. You have to do a few system checks first, which we'll discuss in this chapter.

The short projects in this chapter will help you get your feet wet—don't take that literally, however, when working around electricity—and gain a degree of comfort with your electrical system. Three- and four-way switches require more troubleshooting skills, but we'll cover the most common situations with both switches. We'll also discuss upgrading your current two-wire receptacles and making them safer when the situation calls for it—without updating the entire system with a grounded conductor.

Probing the Problem

You flick the light switch and nothing happens. The coffeemaker, which was set on a timer to go off at 6 A.M., sits with pot of cold water on your kitchen counter. Before assuming that the devices are shot, follow this checklist:

❏ Confirm that the circuit has power and that the fuse hasn't burned out or the circuit breaker tripped.

❏ Check to see if the appliance or fixture is working by checking light bulbs, cords, and plugs.

❏ Inspect the connections at the fixture and at the terminal screws to ensure they are tight.

❏ Check for a problem in the circuit itself.

The very first thing you should do is confirm that power is getting to the device. Every electrician has a story about going on a service call to repair a dead circuit only to discover that the breaker had tripped and no one checked. That's a pretty expensive discovery for a homeowner. Your first step is to examine your fuse box or service panel.

Positively Shocking

Turn the power off before pulling a device out of a box. This isn't necessary if you can test it without removing it, but you can get shocked during removal if it's shorting. Running down to the panel box only adds a few minutes to the job, but it increases your safety immeasurably.

Does everything look okay? Are there burned-out fuses or tripped breakers for that circuit? Check carefully. Some breakers have very little sponginess and don't move much when they trip. You might have to test several breakers if you haven't done a circuit map and are uncertain which breaker controls the failed device.

If you can eliminate the power source as the problem, check the connections (see the following figure) by taking the following steps:

1. Turn the power off.

2. Remove the cover plate from the device, and unscrew the device from the box.

3. See if the terminal screws are tight and have good contact with the wires.

4. If the device is back wired, there shouldn't be any bare wire showing, only insulation.

5. Check the wire nuts or taped-and-soldered connections to be sure they're tight.

If there are no problems with the connections, you'll have to probe further with your handy voltage tester and a continuity tester.

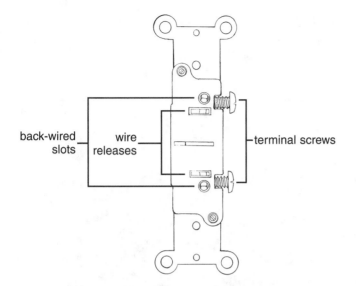

A rear view of a single-pole switch.

back-wired slots — wire releases — terminal screws

Checking the Devices

Switches, appliances, and fixtures test a bit differently than receptacles. The first three are tested for continuity and power and require both testing tools. A continuity tester will indicate whether the circuit's pathway within the switch has any breaks in it from metal fatigue. It also checks other appliances and fixtures for similar breaks. Let's start with testing for a switch.

Follow these steps to test a switch for power:

1. With the switch off, touch one of the voltage tester's probes either to the bare end of the ground wire (the inside of the wire nut holding the neutral wires together) or, if it's a metal box, to the side of the box.

2. Place the other probe against each black wire either at the terminal screw or at the back-wired slot.

3. The bulb should light up for at least one of the hot wires, the line wire coming from the panel or fuse box. If the tester does not light up for either black wire, the problem is somewhere in the circuit between the panel and the device. (Go to step 4 if it does light up.)

4. Turn the switch on and check the other black wire, which is the load conductor. If the bulb on the tester does not light up, the switch is bad and needs to be replaced.

5. If both black wires show current passing through them, recheck the fixture and the appliance because the problem is not with the switch.

6. Note: In some older homes with knob-and-tube wiring, the neutral wire has been switched and used as a "hot" conductor. This makes the task of troubleshooting much more difficult. If you have any questions or concerns while testing, call an electrician.

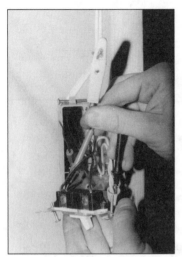 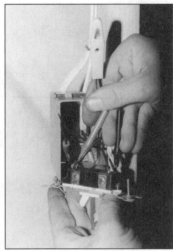 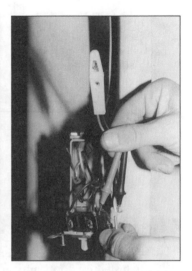

  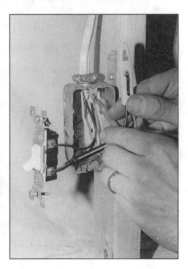

Testing terminals and testing back wiring.

Continuity Coming Up

A continuity test will tell you whether a switch's metal components, which are critical for the flow of the current, are intact or broken. A continuity tester is battery-powered and provides a current that passes from the tester's clip through a device or fixture. The tester's other component is a probe that lights up if a current is passing through the device as designed. To perform the test, a switch or fixture must be disconnected from its power source and removed. (The tester will supply the current for the test.) A continuity tester should never be used on a live current.

The following figures show you how to do a continuity test on a single-pole switch and a three-way switch. If your tests show you that the switch is the problem, it's time to replace it.

If your tests show you that the switch is the problem, it's time to replace it.

Bright Idea

It's almost easier to simply hook up a new single-pole switch to a circuit than to run a continuity test on one that isn't working. By the time you've tested for voltage and have removed it to test for continuity, you could have popped a new switch in and known right away if the switch is the problem.

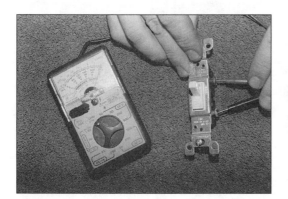 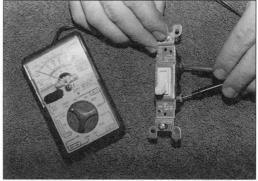

Attach the tester's clip to one of the screw terminals, and attach the probe to the second terminal. Flip the switch to "On" and "Off." If the tester lights up in the "On" position but not in the "Off" position, the switch is good.

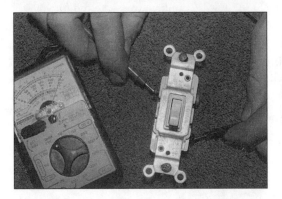

Attach the clip to the common screw terminal (the darker screw), and attach the probe to one of the traveler terminals. The switch is good if the tester lights up when the switch is in one position but not both. Attach the probe to the other traveler terminal and repeat the test, moving the toggle. This time, the tester should light up when the toggle is in the opposite position from the first test with the other traveler terminal.

New Switches

The easiest switch to replace is the single-pole switch. Before popping in a new one—and this is true with any device—read the specifications on the old switch. These usually are listed on the metal mounting strap and include the following:

➤ The amperage and voltage ratings

➤ The type of current it will carry (AC only for house current)

➤ The type of wire that's compatible with the device (CU for copper only, CO/ALR for copper or aluminum, ALR for aluminum only)

➤ Its Underwriters Laboratory or other testing service listing

The back of the device will indicate the acceptable wire gauge and a stripping gauge for measuring the amount of insulation to be removed prior to installation.

154

The location of your switch will determine how its replacement gets wired. A switch can be at either the middle or the end of a run (one complete circuit). These positions in the run are simply defined:

➤ A *middle-of-the-run switch* can be anywhere between the beginning and the end of the circuit. There will be at least two cables entering the box (at a minimum, one on the line side coming from the panel and one leading to a fixture or other device).

➤ In an *end-of-the-run switch* (also referred to as a "switch loop"), the cable runs from the fixture to the switch. This requires special treatment of the white wire.

The following figures show a middle-of-the-run switch and an end-of-the-run switch. Remember to shut the power off at the service panel or fuse box and to test the switch with a voltage tester before removing the wires. Note the condition of the ends of the wires. You don't want to reuse damaged or nicked wires. If you find any damage, cut off the minimum amount of wire necessary to remove this section and then strip off sufficient insulation (about ⅝ of an inch) so the wire will make a solid contact.

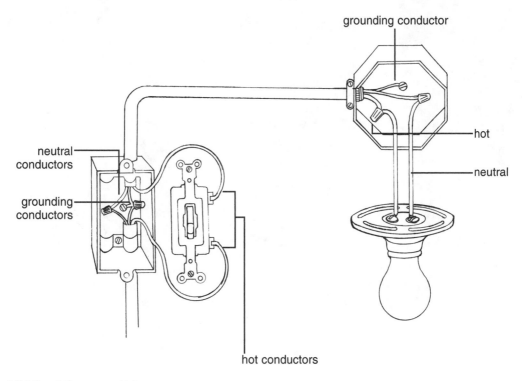

Middle-of-the-run switch.

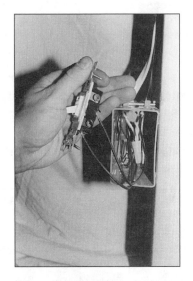
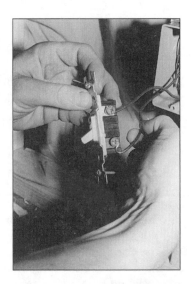
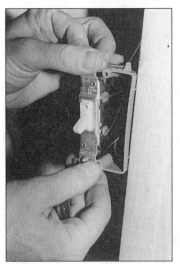
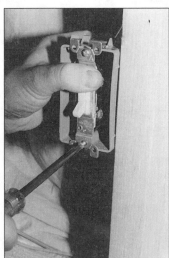

A middle-of-the-run switch: Loosen the terminal screws and carefully remove the black wires. Attach and secure the wires (hand tighten, but don't overtighten) to the new switch and install with the toggle in the "Off" position pointing down. (The black wires can go on either terminal screw.) Carefully tuck the wires back inside the box, pushing them to the side of the neutral wires as you mount the switch to the box. The mounting strap has two $6/32$ screws that insert into the box. Secure the switch to the box, making sure it's straight, and reattach the cover plate.

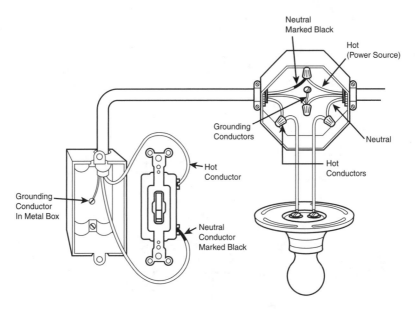

End-of-the-run switch (or switch loop): In this case, only one cable enters the box, and it's the one coming from the fixture. The white wire will serve as a black conductor, as shown in the following figure. The code requires that the white wire be the "hot" or phase conductor leading to the switch. That is, it will connect to the incoming black wire (the line conductor) in the fixture's electrical box. The black wire leading to the switch will be used as the load or switched conductor, which will run back to the fixture. To avoid confusion, you can mark the white wire with a small piece of black electrical tape so anyone looking at the switch in the future will know that both wires are hot.

Three-Way Switch

Replacing a three-way switch is more involved than replacing a single-pole variety. Now you have traveler wires to deal with (these connect the two three-way switches) as well as the common wires. This means the cable running between the switches is 12/3 (or 14/3) cable rather than the more common 12/2 (or 14/2).

Follow the same safety and testing procedures that you would with a single-pole switch. Because three-way switches (and four-way switches) are more expensive than a common single-pole switch, you want to be certain that the switch is really broken before throwing it away and replacing it.

Ask an Electrician

Cable is manufactured according to its gauge and the number of wires it contains. **12/2** is 12 gauge with one hot wire (usually black), one white or gray neutral, and a ground. **12/3** also is 12 gauge with two hot wires (one black and the other commonly red), one neutral, and the ground. 12/3 enables you to run one cable to various nonsequential devices.

A three-way switch; traveler terminals; traveler wires; common wire; common terminal; ground wires with jumper.

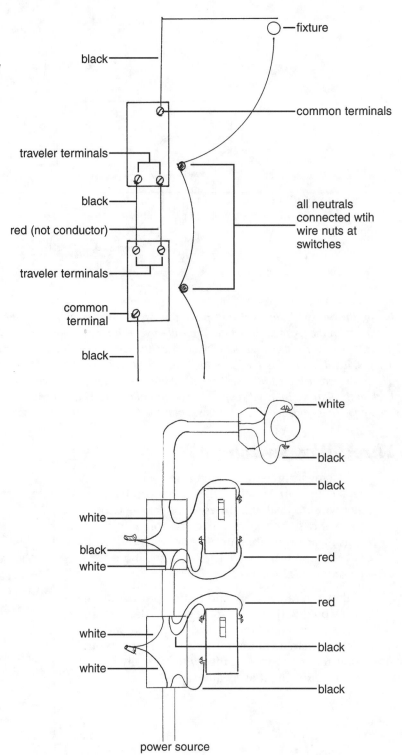

fixture

black

common terminals

traveler terminals

black

red (not conductor)

traveler terminals

common terminal

black

all neutrals connected wtih wire nuts at switches

white

black

black

white

black
white

red

red

white

black

white

black

power source

When removing a three-way switch from its box, note to which terminal screws or back-wired slots the wires are connected. Mark the common wire with a small piece of masking tape, or attach each wire to the new switch as you remove them from the old switch. The common terminal screw usually is copper; the traveler terminals are brass or sometimes silver. Note whether the neutral wire is being used as a hot conductor.

Four-Way Switch

A four-way switch has two sets of traveler wires running between it and a pair of three-way switches. There is no common wire nor is there a common terminal. The continuity test for a four-way switch requires a few extra steps. You need to put the clip on any pair of traveler screw terminals separately and then touch each of the other screws with the probe. This is a total of six tests for each position of the toggle (see the following figures). The test should show two continuous currents for each position of the toggle switch. (The paths between specific traveler screw terminals vary with different manufacturers.)

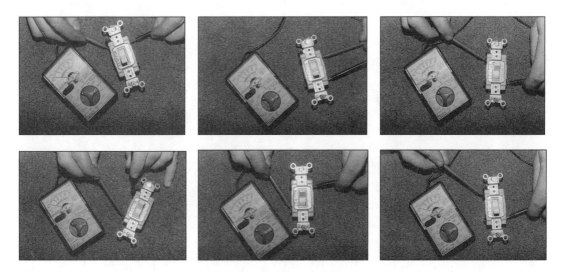

A four-way switch; 1, 2, 3, 4; testing screw terminals by pairs (1-2, 3-4, 1-4, 2-3, 1-3, and 2-4).

A four-way switch box has two cables with three conductors coming into it (thus four hot conductors or wires). Two are black; the other two are a second color, most likely red. When you replace the switch, be sure to match the wires to the correct traveler terminals. New four-way switches either match their terminals up by color (two are brass and two are copper), or the back of the switch might have wiring instructions. This makes your job easier. You simply have to match one color of wire insulation to one set of screws (red wires to brass screws, for example). See the following figures for a typical installation.

A four-way switch.

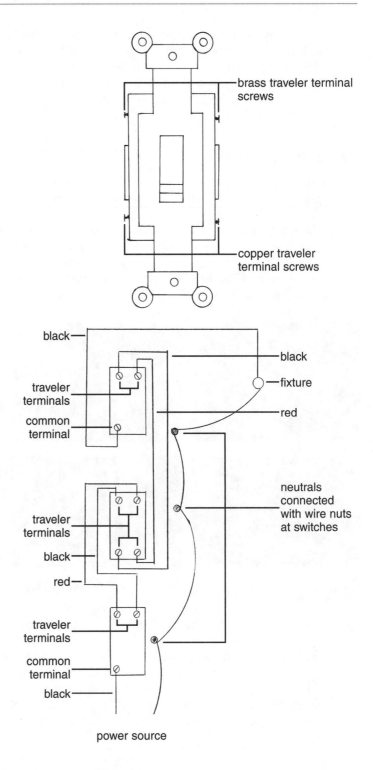

brass traveler terminal screws

copper traveler terminal screws

black

black

fixture

traveler terminals

common terminal

red

traveler terminals

neutrals connected with wire nuts at switches

black

red

traveler terminals

common terminal

black

power source

The Great Outdoors

You cannot replace an outdoor switch with an indoor switch unless you also use a bubble-type cover or a cover with a flip-style lid. These are weatherproof covers. Better yet, you can use a cover with a built-in, horizontal, lever-type switch that comes with a foam gasket between the cover plate and the box. The lever activates a regular toggle switch underneath. Other than that, the replacement procedure is the same as a regular single-pole switch.

Dimmers

You can replace any interior single-pole switch with a dimmer if the box is large enough to accommodate the larger body of the dimmer. Don't try to pack it into a tight or overcrowded box because this is a fire hazard (see the instruction sheet that comes with the dimmer). Dimmer switches come with about four inches of their own wiring or lead wires (line and load and ground wire) ready to connect with cable from the circuit with *wire nuts*.

Old Wire, New Switch

It can be difficult working with the deteriorated knob-and-tube wire ends inside a box. You might have to snip off the end, and the remaining wire can be a little too short to easily connect to a new switch or receptacle. In this case, you can pigtail a short, new piece of wire to the existing wire and connect the pigtail to the terminal screw on the device. This also will bring the wires into compliance with the NEC, which calls for six inches of workable wire length inside a box. The following diagram shows this type of pigtail.

Ask an Electrician

A **wire nut (Wire-Nut Ideal Industries, Inc.)**, also known as a **solderless connector,** is a plastic, twist-on connector used to connect and protect wire ends that have been twisted together. Each size nut is color-coded according to its manufacturer and can only accommodate a certain number of wires.

Knob-and-tube wiring with pigtails.

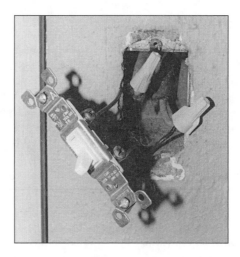

Disreputable Receptacles

Receptacles are pretty long-lasting, but old ones eventually can give out when the clips no longer hold a plug snuggly. There also are drawbacks to some old receptacles if they're neither polarized nor grounded. Receptacle bodies also get broken if furniture or toy trucks somehow bang into them. (This happens at gyms all the time, only barbells do the damage.)

You want your receptacles and their cover plates to be intact. Broken or missing sections can set up you and yours for a shock or worse. When replacing an old receptacle, you can't simply pop a new, grounded receptacle into an existing two-slot outlet. It doesn't work that way, although there is a trick you can do with a GFCI that will give you some protection, but it will not ground any equipment plugged into the receptacle. A GFCI does not give grounding protection unless a grounding conductor already is present.

Check and Check Again

Receptacles, like switches and fixtures, need to be checked with a voltage tester before you do any work on them. The test is similar for both grounded and ungrounded receptacles, except you'll be testing for grounding as well with the former. A grounding test can only be done with the power on.

With the power off, insert both ends of your voltage tester into the slots of the receptacle. The light in the tester should not go on. If it does, the power has not been turned off, or the wrong circuit was shut off. Even if the tester bulb does not light up, you can't be sure that the current is off. The receptacle might be damaged but still receiving a current.

Remove the cover plate and carefully pull the receptacle out. Place one probe on the brass terminal, which should be connected to the black or hot wire. Place the other probe on the silver or neutral terminal. You must touch both terminals to complete the circuit. The bulb shouldn't glow if the power has been shut off.

Positively Shocking

Be sure to install a correctly rated receptacle when you replace an existing one. Most will be 15 amp. Installing a 20-amp receptacle on a 15-amp circuit can lead to overloading and can be a fire hazard.

Testing for grounding.

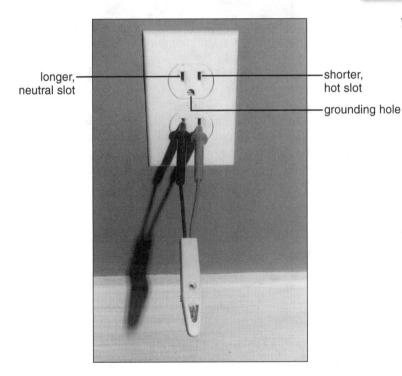

longer, neutral slot

shorter, hot slot

grounding hole

163

Test for Grounding

You can test a grounded, three-slot receptacle for grounding (remember, the power is on) by placing one probe of your tester in the short slot and one in the hole for a plug's grounding pin. The short slot is for the hot wire. The bulb should glow to indicate that the receptacle is grounded. If it doesn't, keep one probe in the grounding hole and place the other one in the longer, neutral slot. In this position, if the bulb glows, it shows that the receptacle is grounded, but the black and white wires have been reversed (they're attached to the wrong terminal screws) and should be corrected. If the bulb doesn't glow in either case, the receptacle isn't grounded.

A three-slot receptacle that isn't grounded is misleading and dangerous to a user. It might indicate that only the individual receptacle is incorrectly wired or that it was inadvertently used to replace an ungrounded receptacle. Either way, you want to know so you can correct the problem.

Testing for grounding; the shorter, hot slot; grounding hole.

Two-Wire Grounding

Do the following to test a two-slot receptacle:

➤ Place one probe in the hot slot and the other end on the screw securing the cover plate. The screw must be clean as well as paint and grease free.

➤ If the receptacle is grounded, the tester's bulb will light up.

➤ Put the probe in the neutral slot if the tester does not light up in the hot slot. If it lights, the receptacle is grounded, but the neutral and hot wires have been reversed and are attached to the wrong terminals. If the bulb doesn't glow at all, the receptacle isn't grounded.

➤ To be absolutely sure that the receptacle is grounded (if your test indicates that it is), turn the power off and remove the cover plate. Check to see if an actual grounding conductor is present.

Testing a two-slot receptacle for grounding.

Installing a New Receptacle

Receptacles are a little more straightforward than three- and four-way switches. With a single duplex receptacle, you're dealing with one or two cables coming into the box. An end-of-the-run receptacle will have one cable, and a middle-of-the-run will have two. The receptacle has two sets of terminal screws, silver for the neutral wires and brass for the hot.

After shutting off the power and testing the terminal screws, remove the outlet by loosening the screws attaching it to the box. Remove the hot and neutral wires, noting their position on the outlet (hot upper, hot lower, neutral upper, neutral lower) by marking the position on an attached piece of masking tape. Reconnect to the new receptacle in the same locations, and gently push the wires back into the box while reattaching the new receptacle. Turn on the power at the service panel or fuse box and test.

Note the wire locations on the receptacle.

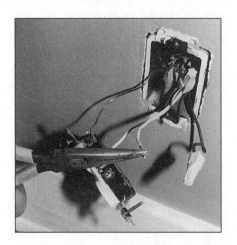

Installing a new receptacle.

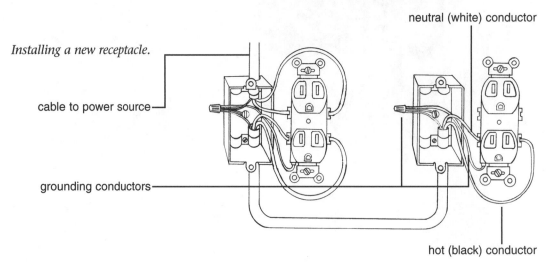

neutral (white) conductor

cable to power source—

grounding conductors—

hot (black) conductor

166

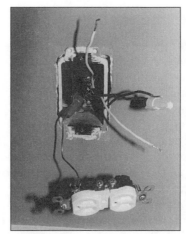

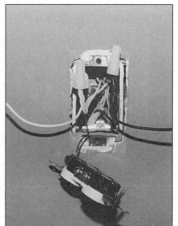

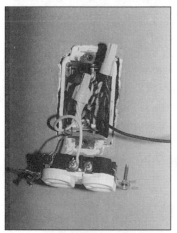

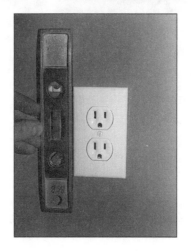

Grounding an Old Receptacle

A properly grounded system ties each device, appliance, and fixture back to the service panel with a separate grounding conductor (the bare copper or green insulated copper wire). It's unrealistic to attempt this with an old electrical system unless you're replacing it (in which case, the grounding would be part of replacing the system).

You also could install a GFCI in place of an existing, two-wire receptacle.

The National Electrical Code allows an ungrounded, two-wire receptacle to be replaced with a GFCI. A GFCI can even protect any receptacles downstream (away from the panel or power source). A GFCI used in this manner will only protect you from ground faults; it will *not* act as a ground for any equipment plugged into the receptacle(s). As a rule, it's best for GFCIs to protect only a single box, not multiple receptacles. If you try to use one GFCI to cover multiple receptacles, you might experience nuisance tripping due to the greater sensitivity to current fluctuations. A GFCI installed to replace a two-wire receptacle should be marked "No Equipment Ground."

A GFCI must be wired according to stamped terminals on the back of the receptacle. They will be marked "Load" and "Line" as well as "Hot" and "White." The hot wire (which runs from the panel or fuse box) is the line conductor; anything going off to another load or receptacle is the load conductor. How do you know which is which? You'll need your voltage tester.

With the power off and the old receptacle removed, separate all the wires in the box so they're not in contact with each other (or with the box if it's metal). Turn the power back on, and put one end of your probe on one hot wire and one on the neutral that is paired with the hot you are testing. If the bulb doesn't light up, try the other black wire and neutral. The one that lights up the tester's bulb is the line conductor. It's the one receiving power from the current you switched back on at the service panel or fuse box. Connect this to the "Line," "Hot" side of the GFCI. It is very important that the line side hot and neutral conductors or wires be connected to the "Line" side of the GFCI; otherwise, the GFCI will trip or will not work at all. If the line and loads are reversed, the GFCI will still have power if it is tripped, producing a hazardous situation.

As an alternative to installing a GFCI to replace an ungrounded receptacle, it is permissible to install a grounding conductor to an ungrounded circuit by using an individual No.12 insulated green copper conductor to connect each receptacle being grounded to the closest cold-water pipe. The grounding conductor will then have to be secured to the pipe using an approved clamping device. It also can be run directly back to the panel and installed in the grounding/neutral bar.

Positively Shocking

Remember, a GFCI is used in places such as bathrooms to control dangerous electricity and possible shock. If you wire a GFCI incorrectly, it will not protect you. There is nothing more dangerous than thinking you're protected when you're not.

A GFCI.

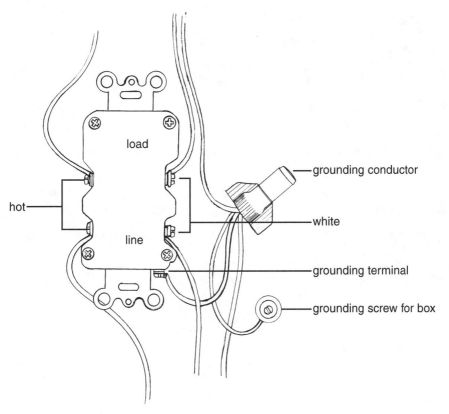

load

grounding conductor

hot

white

line

grounding terminal

grounding screw for box

Aluminum Wiring

Aluminum wiring now requires receptacles and switches marked CO/ALR (see Chapter 6, "When You Buy a House"). Do not use any unmarked devices or devices marked CU or CU-CLAD ONLY. You must wire-brush the connections and apply an antioxidant paste to the ends of the wires before connecting them to the terminals.

Bright Idea

It's easier to install a plug-in GFCI directly into your existing two-wire receptacle than to install a GFCI. Consider this option before replacing your old receptacles.

The Least You Need to Know

➤ Before replacing any switch or receptacle that you think is defective, make sure you don't have a problem somewhere else in the circuit.

➤ Even if you've turned off the power to a device, check it again with a voltage tester.

➤ Make sure the existing wiring pattern is correct for grounding and polarization before repeating it with your new receptacle.

➤ A GFCI can be installed in place of a receptacle without a ground. It will protect you, but it will not ground anything plugged in to the receptacle.

Lighting Up

Before Thomas Edison came up with a working light bulb, we burned different substances to provide us with light. We burned candles, oil, and kerosene until the late 1800s, and in some urban areas, natural gas was used into the 1920s. Using small flames as a source of work and reading light left something to be desired. (The traditional lighting of the Christmas tree occasionally burned down the house.)

The advent of electric lighting changed our lives forever. Workplaces have become safer and so have our homes. We are less dependent on natural lighting, so productivity has increased dramatically. On the downside, because we can now wander around the house at all hours of the day and night and see where we're going, we sleep less than ever before.

Today we have a vast array of lighting options to choose from for our homes. From a basic incandescent lamp to the newest halogen fixtures, we can light up every corner and do so with timed switches, dimmers, or a standard toggle switch from multiple locations. This chapter introduces you to some of these choices and how they can affect your electrical remodeling or additions to your system. There really is life after your vintage 1960s ceiling fixtures, but you might want to hold on to the lava lamps.

How Illuminating

We are way beyond the point when lighting was simply functional, allowing us to work and not stumble around after the sun went down. If function was all it meant to us, every room in our house would have one huge, efficient, fluorescent light fixture on the ceiling and maybe a night-light or two for after dark. Instead, lighting does much more such as …

➤ Create a mood or atmosphere.

➤ Define a space.

➤ Provide security and safety.

➤ Highlight artwork or a section of your home.

Your lighting needs will be defined by these factors and others. Before you install a particular type of lighting, ask yourself the following questions:

➤ Who will be using this area and for what purpose?

➤ Do I want a traditional or modern look?

➤ How often will anyone be in this room?

➤ How much am I willing to spend?

➤ Is energy conservation important to me?

At a minimum, the code calls for one switch-controlled light per habitable room. Hallways, stairways, and garages also must meet this code requirement. This can be accomplished with permanent fixtures, such as ceiling lights, or through a switch-controlled receptacle into which a lamp can be plugged. Bathrooms and kitchens, however, must have an installed fixture. Your first step is to establish your minimum lighting needs and then choose the style of fixture you want to meet them.

Electrical Elaboration

Despite Edison's introduction of the incandescent lamp, it would be years before electrification was prevalent around the country. Builders eventually began installing wiring in homes even if electricity wasn't available in the immediate area. As a backup to unreliable electrical generation, both gas and electric lighting were installed in new homes in some locations until around 1920. Homeowners wanted the assurance of gas if the power went out for an extended period of time.

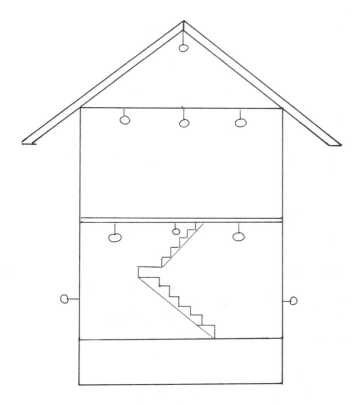

Typical locations of switch-controlled lighting; Required at:

Outdoor entrances.

Basements (all finished rooms, stairs, and storage areas or rooms with HVAC or water heaters which may require maintenance).

First floor: all living space (cannot have switch-controlled receptacle in bathrooms or kitchen).

Second floor: all living space and stairs.

Third floor/attic: if finished or used for storage.

Garage: inside, but not mandated for exterior.

Measuring Your Lighting Needs

The NEC calls for a minimum calculation for general lighting and receptacle loads of three watts per square foot of living space. This comes out to approximately one circuit every 575 square feet. That doesn't amount to a lot, but remember, electrical codes only establish minimum standards. In reality, you'll want lighting everywhere. Consider the different areas of your house and their individual needs:

➤ **Kitchen.** Overhead lighting, natural light from windows and sliding doors, work light over counters, a light over the stove.

➤ **Bathrooms.** Primarily lights over the sink(s) and lights over the bathtub and toilet, depending on the size of the room.

➤ **Dining room.** A hanging light over the table, recessed ceiling lights, or possibly wall sconces. This room often has a dimmer switch to tone down the light.

➤ **Bedrooms.** Children's rooms often get ceiling lights. Master bedrooms might depend more on reading lamps and switch-controlled receptacles, although large rooms can use recessed ceiling lighting as well.

➤ **Hallway.** You might want a long track light to highlight artwork on the walls.

➤ **Garage.** At least one light per bay over the hood of the cars. It's even better to add one or two at the other end so the trunks are illuminated.

➤ **Basement.** Depends on whether the space is finished or unfinished. In either case, you want at least enough ceiling light to cover the entire area thoroughly, leaving no dark spots.

➤ **Closets, storage rooms.** At least one ceiling light.

➤ **Outdoors.** At a minimum, one light over each entry door and over the garage doors. It's even better to consider lights to line walkways and illuminate gardens or security lighting for back and side yards.

Your use of a room obviously will determine your choice of lighting fixtures, their locations, and their number. A single fluorescent ceiling light will fulfill all the working requirements of most closets because the requirements are pretty basic: to shed enough light for you to identify and choose your clothes. A kitchen, on the other hand, requires all kinds of light for a modern homeowner. You need lights over counters for close work so you can chop, dice, and mince vegetables instead of your fingers. Overhead lights enable you to read the newspaper and get a better look at what's hiding in the back of your pantry. A dimmer-controlled hanging light over the eating area lets you tone things down for a late-night meal.

Anywhere you've got a wall, ceiling, or floor, you can install a light. It's simply a matter of extending a circuit or running a new one and choosing your fixtures. The science of lighting is a little more complicated.

Distinguishing a Lumen from Illuminance

Light output is measured in lumens. According to *The American Heritage Dictionary of Science,* a lumen is a unit of luminous flux equal to the amount of light from a source of one candela radiating equally in all directions. A candela is a unit of luminous intensity equal to $\frac{1}{60}$ of the radiating power of one square centimeter of a black body at 1,772°C. You can draw two conclusions from this information:

➤ The higher the lumen measurement, the more light you'll have to work with from a fixture.

➤ Authors can easily get carried away when they have too many reference books at their disposal.

Illuminance, which is measured in foot-candles, is the amount of light hitting a point on a surface. A foot-candle is (easily enough) defined as the amount of light produced by one candle on a surface one foot away. We can't see illuminance, but we do see luminance or brightness, although this is somewhat subjective. (What appears to be dim light to me might be plenty bright to you.) Architects and lighting consultants take all these measurements into consideration when they calculate the lighting needs of buildings.

Comfortable lighting selections and light levels are determined by the tasks that require the lighting, the distance between the light and the task, and the degree of glare. One definition of glare is excessive contrast between the intensity of light on a particular object or surface and the surrounding area or background; indirect glare is the glare produced from a reflective surface. Too much contrast between them causes glare. (Computer screens are a common example.) You can reduce this glare by ...

➤ Installing fixtures that keep the light level appropriate for the task at hand.

➤ Using a louver or a lens to block or redirect the light.

➤ Carefully considering the placement and spacing of light fixtures.

Another measurement of lighting quality is how well it enables you to see colors accurately. The better the color rendering, the more pleasing the living space. Color-rendering capability is based, naturally enough, on the color-rendering index (CRI), which measures from 1 to 100. (Natural daylight measures at 100.) The higher the rating on the CRI, the more lifelike and accurate the object being viewed.

Know Your Lighting

Lighting is defined by its use in our homes and places of work. Designers and architects break it down into several categories:

➤ **Accent lighting** emphasizes or highlights a specific area or object and directs our attention to it.

➤ **Ambient lighting** is general illumination.

➤ **Task lighting** is for illuminating work and tasks.

Bright Idea

When determining your lighting needs, keep in mind how reflective the various surfaces in a room will be and how much daylight is available. Dark-green carpet won't reflect light the way white vinyl flooring and countertops will. Too much light is just as much a problem as too little light.

Positively Shocking

Never use a higher-wattage light bulb than a light fixture can handle. This can lead to overheating and possibly a fire. Instead, install fixtures that can handle a higher wattage than you need so you have the choice of using either the higher- or lower-wattage lamps.

It's never a bad idea to install plenty of ambient lighting, even if you later decide it's more than you immediately need. At some point in the future, you might move things around and decide you need more lighting. I wouldn't recommend tearing up

175

the walls just to install fixtures, but if you have an open ceiling or already are doing some installations, consider a few extra light fixtures if the circuit permits.

If you walk into a lighting store or the lighting section of a home-improvement center, you'll see dozens and dozens of fixtures to choose from. Where do you start?

Aim High, Low, and Wide

Light from a lamp is aimed somewhere, whether it's the top of your desk or your workbench. Even general ambient lighting gets directed somewhere. Recessed ceiling lights and adjustable spotlights can provide as broad or as focused a beam of light as you desire. Some lights are installed as wall washers, meaning they shine down a wall either to highlight artwork or other collections or simply to draw your attention to the perimeter of the room, conveying a greater sense of size than might truly exist. The advantages of recessed ceiling lights are their versatility and unobtrusiveness. Let's face it, a chandelier automatically draws attention to itself—especially if one of your party guests is swinging on it. A recessed fixture is far more subtle and almost hides in the background.

Some fixtures can serve more than one purpose. A wall sconce, for example, can serve general, task, and accent lighting needs. This versatility is a huge advantage over ceiling lights when you're remodeling because it's far easier to wire and install a wall fixture than to install most ceiling fixtures.

Bright Idea

For bedroom reading lights, consider a halogen lamp by the bedside. This will give out a very focused, narrow light rather than filling the room the way an incandescent lamp will. A narrow light will be less disturbing to a sleeping spouse.

Lighting Up Outside

I think exterior lighting is always a plus with any home (see Chapter 21, "The Great Outdoors"). Good lighting will welcome you and your guests on a rainy night, provide some measure of security for your family, and illuminate address numbers, door locks, and staircases. Before you decide to install fixtures as powerful as Batman's searchlight, consider the following:

➤ Know the size of the fixture and its scale compared to your house.

➤ Think about the location and aim of the lights and their effect on your neighbors. (A little light goes a long way at night.)

➤ Caulk the top seam between the fixture and the section of the house where it's attached to ensure that water stays out. Leave the bottom uncaulked so that, if moisture does get in, it has a place to exit.

➤ Think twice before installing solid-brass fixtures. They won't rust, but eventually most will tarnish and need polishing.

Outdoor path light.

(Progress Lighting)

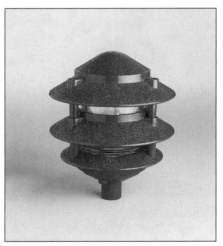

Outdoor pagoda light.

(Progress Lighting)

177

Electrical Elaboration

In some instances, outdoor lighting can be dangerous to wildlife. In parts of Florida and other coastal areas, residents and businesses are asked to turn off their bright lights during certain cycles of a full moon because the lights can confuse sea turtles who see them as a signal to come ashore and lay their eggs.

Installing outside lighting, like many tasks, can be done the easy way or the hard way. The easy way means mounting all the fixtures on the walls of your house (and porch ceilings), which means you can pull your wires from inside the house. The hard way means digging ditches and running wires and conduit underground, although this will give your yard a much more dramatic presentation. As a final consideration, think about what your outdoor lighting will look like from *inside* the house. You can enjoy your yard even in the winter if you set up lighting that accents it well.

Combining Lighting Styles

Most general living space will accommodate more than one lighting style. A closet obviously doesn't need accent lighting unless you make a point of giving your guests a tour of your shoe collection. A dining room needs ambient light, but it can become very dramatic with, say, floor-to-ceiling lights illuminating the side walls while the recessed ceiling lights are off and a few lit candles are on the table. A mix of lighting offers the most options and can present a room and its occupants at their best in a variety of settings.

Whether you're building something new or remodeling, keep your furniture in mind rather than strictly installing lights by formula (so many per square feet at such and such a distance from each other). You might have a grand piano ready to nestle in a corner of the living room or a windowless wall just waiting for your collection of family portraits. Either situation calls for very specific fixture placement.

Close-to-ceiling incandescent.

(Progress Lighting)

Close-to-ceiling compact fluorescent.

(Progress Lighting)

Sconce.

(Progress Lighting)

Track lighting.

(Progress Lighting)

Semi-flush style.

(Progress Lighting)

Chandelier.

(Progress Lighting)

Looks Are Something

The cheapest porcelain fixtures with 100-watt bulbs might provide safe lighting, but they won't be much to look at. Visualize the appearance of the fixture and the lamp as well as their function. Some fixtures literally are works of art (Tiffany lamps, for example); others are designer-created and are very striking to look at. Do you want to look at a brass hanging light over the kitchen table or cobalt-blue steel? You're going to be looking at them every day, so take your time choosing your fixtures.

What's Your Type?

Lighting is divided by the type of lamp used and the style of the fixture. Lamp types include …

➤ Incandescent.

➤ Fluorescent.

➤ Halogen.

Each of these has distinguishing characteristics, as described in the following list. When it comes to fixture styles, the sky's the limit. They range from antique reproductions to one-of-a-kind works of art (with prices to match). The following are some of the most common light fixtures:

➤ Flush-mounted ceiling lights that include square, mushroom, or round domes

➤ Hanging ceiling lights and chandeliers

➤ Surface fluorescent lights

➤ Recessed ceiling lights

➤ Track lighting

➤ Bath bars

➤ Sconces and wall-mounted lights

➤ Wall washes

➤ Undercabinet-mounted lights

Bright Idea

Consider installing adjustable recessed fixtures so the light beam can be aimed in different directions as your decorating changes. This is less expensive than trying to install additional fixtures.

Ask an Electrician

Although we commonly use the term "light bulb," the more accurate term is **lamp.** A **bulb,** as my technical editor has pointed out, is something you plant in the ground. The complete lighting unit (lamp, lamp socket, housing, reflective material, lenses, and, when called for, ballast) is called a **luminaire.** I use the term "lamp" where possible in this book. Where I must, I use the word "light bulb" for clarity.

➤ Outdoor lights (floodlighting, landscape lighting, pole-mounted lights, wall-mounted lights, and security lighting)

All of these will light up a given area. You just have to decide whether they will provide light that you find both appropriate and pleasing to the eye. Your budget also is a consideration, especially if you're buying fixtures for a major remodel. An outdoor landscaping light, for example, can be a simple pagoda light or an ornate—and expensive—leaded-glass lamp. (The latter is not recommended if you have kids, dogs, or errant adults running around the yard.)

Incandescent

This is the most familiar type of lamp. An electric current passes through and heats a tungsten filament, producing a glowing light. (The term "incandescent" literally means "to glow or become hot.") Over time, heat evaporates the tungsten, and it eventually weakens and breaks. The lamp contains a chemically inert gas that allows the tungsten vapor from the heated element to deposit on the sides of the glass. This is why standard incandescent lamps gradually become darker over time. If they are too small, the tungsten coating would turn them opaque, and they'd be useless as a source of light.

Incandescent lamps are cheap to produce and are versatile in application, but they are considered to be impractical by energy conservationists as a source of light given modern alternatives. They produce a considerable amount of waste heat for the amount of current they draw, and they have a useful life of 750 to 2,500 hours depending on the lamp. Builders traditionally install incandescent fixtures because both they and the lamps are inexpensive and are not likely to meet any resistance from price-conscious buyers. When buying lamps, take note of the voltage rating of the lamp. Typical lamps sold in stores are rated at 115 or 120 volts. At professional lighting stores, you should be able to find longer-lasting lamps rated at 130 volts.

Tungsten-Halogen Lamps

These lamps (which are smaller in size than standard incandescent lamps) also heat up a tungsten filament, but they contain halogen gas. The gas combines with the evaporated tungsten to create tungsten halide gas that deposits the tungsten back onto the filament, extending its life. After the deposit, halogen gas is released and the process starts all over. The smaller size of the lamp enables the filament to heat up to a higher temperature and a higher efficiency.

Because none of the tungsten is deposited on the glass, a halogen lamp burns brighter and has a very focused, intense light.

Electrical Elaboration

Halogen lamps operate at high temperatures. Halogen lights designed for installation under kitchen cabinets can be tested to UL standard 153. This is a safety standard that says the wood above the lamp must have a temperature lower than 194°F or 90°C with the lamp on. Halogen lamps used inside china cabinets (or other cabinets used for display) also must pass this test to get this UL rating.

Fluorescent Lighting

Fluorescent lamps are considered to be the most energy efficient, but they often suffer from a reputation as flickering, eerie sources of light suitable only for institutional settings. They are the light source of choice in industrial and commercial settings because of their efficiency and long life—something worth considering for your home as well. Modern fluorescent fixtures have a place in residential settings.

A fluorescent lamp is constructed with …

➤ A glass tube.

➤ Argon or argon-krypton gas and a small amount of mercury.

➤ *Phosphor* coating on the inside of the tube.

➤ Electrodes at each end of the tube.

As electricity passes between the tube's electrodes, it jostles the mercury atoms, which then give off ultraviolet radiation. The radiation is converted to light when it interacts with the phosphors lining the tube. The fixture itself comes with a ballast to kick-start the current passing within the tube and to keep it regulated. The range of phosphors available to manufacturers enables them to produce lamps with different color tones for different applications. Fluorescent fixtures also produce less heat and more light for the amount of electricity they consume as compared to incandescent lamps.

Ask an Electrician

Phosphor is a phosphorescent substance. **Phosphorescence** refers to the light given off when a substance absorbs certain types of rays such as ultraviolet rays.

One of the biggest changes in fluorescent technology is the shape and size of the lamps. In the past, you were stuck with straight, U-shaped, or circular figures. (The last always seemed to be used outside small-town, drive-in, ice-cream stands for some reason.) Now we have compact fluorescent lamps that can be an efficient substitute for incandescent lamps. A 40-watt compact fluorescent lamp, for example, can replace a 150-watt incandescent lamp and can last up to 10 times longer. (It had better, given the typical cost of $20 or more.) Two types of replacement units are available:

1. Integral units, which include a compact fluorescent lamp and ballast in a self-contained unit

2. Modular units, in which the bulb is replaceable

Now you know that you can replace your incandescent lamps with compact fluorescent lamps, but is it worth it?

When Cheap Power Reigns

At one point in the 1980s, it was estimated that the cost of residential electricity in Seattle was one tenth the cost in New York City. If it's any consolation, our delicatessens weren't anything to write home about. The cost differential isn't that great anymore, but we're still below the national average. Lower costs aren't necessarily a justification for excessive use of electricity, but they will determine whether more efficient fluorescent lamps will ever pay off for you.

Basically, the higher your electricity costs run beyond the national average (around 8¢ per KWH), the more cost-effective fluorescent lights will be in your home. This doesn't mean they will work well for all your lighting needs from a cost standpoint. The longer a light is continually on, the better a candidate it is for a fluorescent lamp. An occasionally used attic or storage-room light is best left with an incandescent lamp. Some fixtures, such as recessed ceiling lights, might not have room for a fluorescent lamp.

Electrical Elaboration

Keep in mind that adding windows also will affect your overall energy costs, probably even more than the amount of light you'll be saving by using natural daylight versus electric light. Nevertheless, it's hard to put a price on the advantages and the appeal of natural light, especially during the winter months. Look at all factors before ruling out additional windows.

Other Considerations

The most comfortable light in many instances is natural sunlight. (You may think differently if you live in the Sahara desert.) If you're remodeling or adding on to your house, think about adding more windows and skylights. Millwork companies can custom match any existing wood window or come close enough with stock material. There are enough vinyl and aluminum window manufacturers around that you should be able to find one that will look like part of your house.

Paint color also affects the impact of light, both natural and electric. Light colors will be the most reflective, but they might not be your first choice in certain rooms. Balance out your color choice with adequate lighting.

Finally, look at your choice of controls or switches. Dimmers are inexpensive, and they greatly expand your options in any room. One minute your living room is washed in light for your Scrabble club's monthly game; the next minute it's dimmed way low for you and your jo. (Scottish for "sweetheart," this word works great on a triple-word score.)

The Least You Need to Know

➤ Good lighting is more than just functional; it provides atmosphere, helps in your work, and adds highlights.

➤ Be sure to match your light fixtures and lamp choices with the lighting requirements you're trying to meet.

➤ It's hard to go wrong installing plenty of ambient or general-purpose lighting.

➤ Outdoor lighting should get just as much thought and attention as your indoor plans.

➤ Fluorescent technology has greatly improved over the years and is a good choice for residential lighting.

Light Fixes

Changing a light fixture can be more involved than simply replacing a switch or a receptacle. Switches and receptacles almost always are housed in electrical boxes, but this isn't always true for light fixtures. If the system is old or has been hacked at enough, you can disassemble an old ceiling light only to find a couple of wires dangling through the plaster without the hint of a box. As you should know by now, this is a dangerous situation because all wire connections must take place within a box. You might feel like cheating by continuing the status quo, but don't. You'll need to install a new box (unless one's already there).

Lights usually get replaced because tastes change. Old fixtures don't often wear out, since they have no moving parts (unlike a switch). Historic or not, the original hanging lights in your Craftsman home might be ugly to your eyes, or you might want to replace more modern fixtures with replication period fixtures to restore your home closer to its original condition.

The usual safety precautions apply to replacing fixtures that apply to any other electrical work—turning the power off is number one—but now you'll sometimes be working off a ladder. For that matter, two of you might be working off two ladders if you have to remove an especially heavy or delicate fixture such as a chandelier. One thing is for certain: With the huge selection of new fixtures to choose from, you're bound to find a replacement that will dress up any room in your house.

Inspect First

There are two main reasons for replacing a light fixture:

➤ It isn't working, and you believe it is somehow broken.

➤ You want to install an updated style or a fixture that will offer more light.

You should do a number of checks before pronouncing a light fixture broken or beyond repair. You already know about checking the switch and the fixture itself for power. In addition, you should look at the following:

➤ The lamp (light bulb)

➤ The socket

➤ The wire connections inside the box

Checking the light bulb is the obvious first course of action—replacing the bulb with one that's working. If it got jostled around in a storage drawer or even on the way home from the hardware store, there's always a chance a new bulb isn't working, so check it in a fixture or lamp you know is working. The next thing to check is the socket. At the bottom of the socket is a small metal tab that makes contact with the bottom of the lamp. Turn the power off and check with your voltage tester by placing one probe on the metal tab and one on the inside of the metal socket. The bulb should not glow. If it does, the power is still on, and you need to shut off the correct circuit. With the correct circuit shut off, test the fixture again with your voltage tester.

With the power off, pull the end of the tab up a little bit using the end of a screwdriver. Screw in the lamp, turn on the power, and try the fixture again. These tabs sometimes become depressed or flattened out and don't form a tight contact with the lamp.

Why would the contact suddenly be broken? All it takes is a slight vibration in the fixture from, say, a large truck passing by. If a light bulb that you know is good doesn't work, you have a problem with the socket. Remove the fixture to test the socket by following these steps:

1. With all glass globes, lampshades, and light bulbs removed, unscrew the fixture from the box by turning the mounting screws counterclockwise.

2. Carefully pull down the fixture and let it rest on top of the ladder. This is critical if it's a heavy fixture.

3. Disconnect the wires from the terminal screws and take down the fixture.

4. Attach the continuity tester's clip to the hot wire terminal screw, and place the probe against the metal tab in the socket. If the tester does not glow, the socket needs to be replaced.

5. Attach the continuity tester's clip to the neutral terminal and the probe to the threaded portion of the socket. Again, if the tester's bulb does not glow, the socket needs replacement.

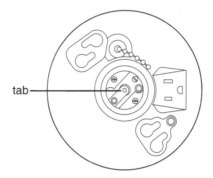

tab

Pull up lightly on tab for better contact with bulb.

(Courtesy of Leviton)

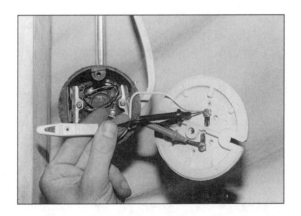

A hot screw terminal; neutral screw terminal; a clip; a probe; a metal tab; a socket.

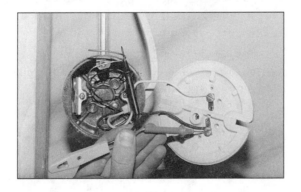

Some sockets are permanently attached to the fixture, in which case the entire fixture must be replaced. Others are attached to the fixture with screws and can be removed and replaced. Take your old socket to the hardware, lighting, or electrical-supply store and purchase an identical replacement.

Bright Idea

Be sure to check for tight wire connections when you inspect your fixtures. Remove the wire nuts (or any tape) to confirm that the wires are making a solid contact. (If not, that's all you have to fix.) It's sure less work than replacing a fixture that doesn't need replacement.

Positively Shocking

Do not install a fixture of higher wattage unless the circuit can support it. Only install a lamp that *is* equal to or less than the wattage recommended on the fixture. (There will be a tag listing the wattage and voltage of the fixture.) A lamp of higher wattage will develop too much heat, which can damage the fixture and the wiring.

The same vibrations also can cause the wire connections to come loose just enough to lose contact with the fixture.

Time to Replace

Installing new fixtures is a doable homeowner project. It's easier to do with modern wiring and boxes, but it still can be done with older types of wire as well. To replace a fixture, follow all safety precautions, read the instructions and diagrams that come with the fixture, and …

1. Turn the power off and test to make sure it's off.

2. Follow the preceding steps for testing the fixture's socket to remove the fixture and undo the wire connections.

3. If the fixture does not have an electrical box, install one (see the following section, "Installing a Box").

4. Install the mounting strap that comes with your new fixture to the box. (The strap, also called a mounting yoke, has predrilled holes set to the dimensions of the fixture.)

5. Connect the black wire and the white wire from the fixture to their counterparts in the circuit wires.

6. Connect the grounding wire to the grounding screw on the mounting strap and to the grounding conductor that might come attached to the fixture.

7. Install a light bulb, turn on the power, and test the fixture. After the test, turn off the power and remove the bulb.

8. Attach the fixture with its mounting screws to the mounting strap.

9. Install the light bulb and the globe.

Installing a Box

A self-supporting retrofit box can be installed in an existing ceiling or wall if your light fixture doesn't have a box. This regularly will be the case with very old wiring or poorly done additions to your electrical system. You'd be surprised how many old incandescent fixtures are attached directly to plaster lath instead of to any kind of box.

Retrofit boxes come in two flavors: metal and plastic. Each is designed to fit snugly against either plaster or drywall by using adjustable ears and brackets that expand and/or tighten against the wall. A plastic box has an attached, U-shaped bracket that tightens like a toggle bolt as its attachment screw is tightened. A metal retrofit box comes with brackets or supports (known as "Madison Holdits" and sometimes as "battleships") that fit between the box and the wall. As they are pulled out, they firm up the fit of the box. The arms of the supports are then bent over the edge of the box, tucked inside, and pinched tightly with pliers.

Another version of a metal retrofit box features a screw-operated support on each side of the box. As the screws are tightened, the metal support wedges the box in tightly against the plaster or drywall. A retrofit plastic box has plastic or metal internal cable clamps that help secure the cables to the box should it ever slip from the opening. Metal boxes are a bit trickier to use if you're unfamiliar with them, so consider using plastic retrofit boxes for your work.

Neither plastic nor metal retrofit boxes are supported by strong attachments to studs or floor joist; they are supported only by brackets maintaining a taut fit (and maybe the hope that no one will yank on them too much). They are not designed to hold heavy light fixtures or ceiling fans.

To install a box ...

1. *Shut the power off* and remove the existing fixture.

2. Carefully push the wires up and out of the way.

3. Using the back of the retrofit box as a template, draw an outline on the ceiling or wall of the hole you need to cut. Make sure it's smaller than the ears on the box, which will brace against the plaster.

4. Carefully cut the hole with a hole saw or a keyhole saw.

5. Take a screwdriver and, while tapping it with a hammer, punch out one knockout for each cable that will be brought into the box. (Move the screwdriver around to broaden the hole and to clean up any sharp edges.)

Bright Idea

Always put a large drop cloth or a piece of plastic under your work area when you're cutting through plaster or drywall. The last thing you want is damage to your hardwood floors from falling chunks of plaster. You also don't want to have to vacuum plaster out of your carpet.

191

6. Bring the old cable or wires into the box.

7. Tighten the mounting bolts on the box. This will tighten the box to the plaster or drywall. Connect the wires to the light fixture and tighten the fixture to the box.

An old fixture installed without an electrical box.

Tighten the clamps to secure the box to the ceiling; in plaster, screw plaster ears to the lath as well.

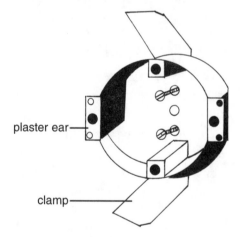

plaster ear

clamp

Cutting a hole in plaster or drywall using a keyhole saw is a chore, especially if you have to saw through lath. Most lath in residential construction is wood, but metal lath is sometimes used in expensive homes and requires either tin snips or a hole saw. (The lath will dull the saw's teeth some.) To get a start on your sawing after you've outlined your hole, drill some starter holes in several locations with your electric drill and a twist bit. For that matter, you can drill a lot of holes all along the outline until you've just about cut out the entire hole. I'm not a purist about this stuff, and I recommend doing whatever you're comfortable with as long as it's safe and works for you.

You have to be careful when cutting through wood lath and be certain your saw blade is sharp. A dull blade and jerky sawing is a great way to pull on the lath, causing it to crack the adjoining plaster. Take your time. Drill more holes if necessary. Don't worry that you're not doing it as fast as an electrician.

Electrical Elaboration

Wood lath is nailed to each stud in a wall. If you saw through it too aggressively with a keyhole saw or a reciprocating saw, it begins to bounce around and crack the adjoining plaster. With the plaster removed for the box, cut down one side of the lath, leaving an inch or so uncut. Cut the other side completely, and then return to your first cut. This way, the lath stays rigid until both cuts are complete. Use a scroll blade for your reciprocating saw. If sawing by hand, use a fine-tooth blade or even a hacksaw blade, which is available with a mini handle that would allow this kind of cutting.

Try a New Style

Styles and fashions change, a fact that's certainly not lost on America's retailers. The dim ceiling fixtures with the square-shaped glass domes from the 1950s and 1960s just don't cut it any more, unless you're into retro-hip lifestyles complete with hula hoops and strange-looking dinette sets. The array of fixture choices today is astounding. Just about any period of fixture can be matched, or a completely updated style can be installed. The only limits are your imagination and your checkbook.

Some fixtures can greatly increase the amount of lighting in a room. A single overhead fixture in a long hallway, for example, might be replaced with track lighting running the length of the hallway, turning it into a great space to display paintings or photographs. As always, make sure your circuit can support the increased current demand should you replace a fixture with one of a higher wattage rating.

There's no need to limit yourself to using incandescent fixtures to replace your existing ones. Fluorescent fixtures, especially compact models, should be considered as well.

193

Fluorescent Fun

Fluorescent lamps last for years, depending on their use, but the components of the fixture can cause some problems. The longer the light is left on, the longer the lamp will last. The usual estimates of a long lamp life are based on three hours of use for every startup of the fixture. The more you turn it on and off for short periods of use, the sooner the bulbs and components will give out.

Fluorescent technology has improved since those wretched, buzzing, industrial-looking lights were installed in the 1960s. If your lights are that old, consider replacing the entire fixture before you replace a malfunctioning ballast.

Positively Shocking

Article 410 of the National Electrical Code spells out the rules for installing light fixtures. Fluorescent fixtures produce heat, which affects the life of the ballast. Read the manufacturer's instructions before installing, especially in a ceiling that's insulated. This will help you get a long life out of your fluorescent fixture.

The Pieces Inside

All fluorescent fixtures have the following major components:

➤ Ballast

➤ Sockets

➤ Lamps (a.k.a. light bulbs or tubes)

The ballast is like a small transformer inside the fixture. It has two major jobs:

➤ First, it provides the spark that sets off the gas within the tube, making that gas fluoresce. This requires higher voltage.

➤ Then, the transformer reduces the voltage to the (very low) level necessary to sustain lighting.

There are three main types of ballasts or fluorescent circuits: preheat, rapid start, and instant start. (The type is identified on the ballast casing.)

Preheat technology dates back to the original fluorescent fixtures, and it is used today mostly for certain low-wattage fixtures such as compact fluorescent fixtures. This type of circuit uses a starter.

Rapid starts are the most frequently used ballasts today. They maintain a continuous low-wattage circuit to the lamp's filaments so they start up faster (in less than a second). This style of ballast also comes in a version that allows dimming.

An instant-start ballast, as the name implies, ignites the gas in the lamp instantly. It supplies a higher starting voltage than the other types of ballasts, and like the rapid-start variety, it requires no separate starter. The drawback to this voltage boost is that it requires a special lamp. The lamp also will have a shorter life than those used for rapid-start circuits. One way you can distinguish between the lamps is by the number of pins on the ends of the lamp. (These pins insert into the fixture's sockets or tube-holders.) Rapid-start lamps are bi-pin (they have two pins); most, but not all, instant-start versions have only one.

Even More Efficiency

I have already noted that fluorescent fixtures are more efficient than incandescent fixtures in a number of common applications. They can be made even more efficient with the use of an electronic ballast, which uses semiconductor components for greater efficiency and produces no fluorescent "humming." The electronic model costs more than a magnetic ballast, the other ballast type, but energy savings over the life of the ballast should more than recover the price difference.

Fluorescent Woes

It doesn't take much power of observation to figure out whether your fluorescent fixture is having operational problems: It either lights up or it doesn't. It also might flicker, hum excessively, or excrete an ominous-looking black substance near the ballast. Some of these problems can easily be solved; others are a bit more problematic, as shown in the following table.

Positively Shocking

Ignoring the condition of your fluorescent lamps can shorten the life of the ballast. When only one lamp out of two is functioning, the ballast will heat up more than usual and will deteriorate sooner. The same is true if one end of a lamp has turned very black. You also should replace flickering lights and those in which both ends are blackened.

Troubleshooting Fluorescent Lights

Problem:	Lamp does not light.
Cause:	No power to switch on fixture; lamp is loose or dead; problem with sockets, starter, or ballast.
Solution:	(1) Check switch and fixture for power and continuity; check wire connections. (2) Turn lamp off and make sure it's tight in the sockets. (3) Replace one or more of the following: lamp, sockets, starter, and/or ballast. Lamps need replacing if the pins are bent or broken or if the ends of the lamp are blackened. Sockets need replacing if they are chipped or do not allow a lamp to seat completely. A ballast needs replacing if it's oozing oily, black resin or if everything else checks out okay.

continues

continued

Troubleshooting Fluorescent Lights

Problem:	Lamp flickers or will not light completely.
Cause:	Lamp is not seated completely in sockets; starter or ballast is faulty; room is too cold; lamp is dirty.
Solution:	(1) Make sure the lamp is seated properly and the pins are not bent; replace the lamp if damaged. (2) Replace sockets if they're damaged. (3) Replace the starter or ballast if necessary. (4) Fluorescent lamps don't do well in the cold (roughly 50°F or less). If your room is this cold, install a ballast rated for colder temperatures. (5) Remove the lamp and carefully clean it with glass cleaner, allowing it to dry completely before reinstalling.
Problem:	Excessive humming.
Cause:	Ballast is faulty or loose.
Solution:	(1) Remount the ballast; tighten the mounting bolts and the fixture's mounting screws. (2) Replace the ballast.

The simplest maintenance for a fluorescent fixture is changing the lamp. It's a good idea to turn the power off at the service panel first. Most fluorescent lamps are covered with a diffuser, which is a translucent plastic cover installed to soften the light. Diffusers usually can be removed simply by pressing the sides slightly inward (these are held in by a friction fit); others are held in with pins, clips, or end plates. With the power off, follow these steps when replacing a tube-shaped lamp:

1. Remove the diffuser and set it aside.

2. Rotate the lamp one-quarter turn in either direction, and slip the pins out of the sockets. Check the condition of the sockets.

3. Install a lamp with the same wattage rating as the burned-out lamp (check the ballast to confirm the rating) by rotating it one-quarter turn in the sockets.

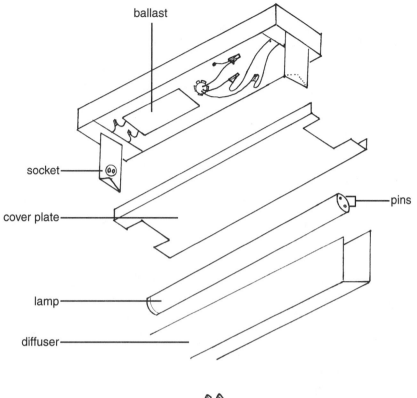

ballast

socket

cover plate

pins

lamp

diffuser

Electrical Elaboration

Disposing of fluorescent components isn't always as simple as tossing them in the trash. The lamps contain small amounts of mercury and should not be broken in a trashcan. Talk to your hauling service about how best to handle them according to local regulations. According to the EPA Greenlights Program (December 1992), some recyclers will disassemble the fixture, removing the ballast and the chamber containing the capacitor.

If you have an older fluorescent light and the lamps don't have darkened ends near the pins, consider replacing the starter before replacing the lamps. Follow these steps to replace a starter:

1. Turn off the power to the fixture.

2. Push the starter in, turn it counterclockwise, and then pull it out.

3. Take the starter to your hardware store and buy a matching replacement.

4. Insert the replacement into the fixture and turn clockwise.

Next on our list of culprits are the sockets. As usual, you start this repair with *the power shut off*. To examine the sockets, follow these steps:

1. With the diffuser, lamps, and cover plate removed, take your voltage tester and make sure the power to the fixture is off. Place one probe on the grounding screw and the other inside each wire nut. If the tester light glows, the power is still on, and you need to turn off the correct circuit.

2. Examine the sockets and remove any that are damaged. Some sockets are held in with screws; some slide in and out.

3. The socket wires will have to be removed. They will be attached either with screws or at push-in terminals. They also might be permanently attached and have to be cut.

4. Take the socket to your hardware or lighting store and purchase an identical re-placement. If the new socket comes with its own wires attached, these will have to be connected to the ballast with wire nuts.

5. Reattach the socket(s), the cover plate, the lamps, and the diffuser.

6. Turn the power on and test the fixture.

The only major component you might repair is the ballast. Before you do that, how-ever, consider the cost and the age of the fixture. You might be better off simply re-placing it entirely with a new, modern fixture.

Follow these steps to replace a ballast:

1. Shut off the power to the fixture and remove the diffuser, lamps, cover plate, and sockets.

2. Disconnect the wires attached to the sockets by cutting the wires halfway be-tween the sockets and the ballast.

3. Remove the ballast (don't let it fall out) and purchase a new one with the same ratings as the original.

4. Attach the new ballast in the same direction as the old one.

5. Reattach the wires from the ballast to the sockets. (Be sure to match the same colors under the same wire nuts.)

6. Secure the sockets, cover plate, lamps, and diffuser.

7. Turn the power on and test the fixture.

Replacing the entire fixture isn't much more work than replacing the ballast, but two sets of hands usually are better than one, given the awkward dimensions of some fixtures. Keep in mind that a replacement fixture should not exceed the load limits for the circuit. Hire an electrician to do this if you're at all uncertain about doing it yourself.

Follow these steps to replace an existing fluorescent fixture:

1. Shut the power off.

2. Remove the diffuser, lamps, and cover plate.

3. After testing for a current with your voltage tester, disconnect the neutral, hot, and grounding wires from the fixture, removing all wire nuts and cable clamps.

4. Carefully unscrew the fixture from the ceiling or wall and remove it. (This is where the extra hands come in.)

5. With your new fixture, carefully thread the circuit wires through the knockout opening on the back of the fixture, insert and tighten the mounting screws, and reconnect the circuit and fixture wires with wire nuts. If there's a cable clamp holding on to the circuit cable, tighten it.

6. Install the cover plate, lamps, and diffuser.

7. Turn on the power and test the fixture.

Lamps

Lamps are just portable light fixtures. They operate using the same principles (a current comes into a load, a bulb lights up, the current exits through the neutral wire), but they have some slightly different maintenance issues. A lamp's failings occur around its cord, plug, and lamp socket.

It's easy to ignore lamp maintenance. After all, how often do we check cords and plugs? Lamp cords consist of No.16 stranded wire that is easily broken if the cord is abused or misused. Plugs can only take a certain amount of yanking and pulling before they give out. Parts are available, but you'll have to decide whether it's worth your while to install a new cord in that $5 desk lamp that you picked up at a garage sale when you were a freshman in college.

The Easy Repairs

You turn on your trusty reading lamp and—nothing. Chances are it's the light bulb. Remove the old bulb, screw in a new one, and you should be ready to go. If not, you'll have to inspect further. Your problem could be ...

➤ The power supply.

➤ The receptacle.

➤ The contact tab at the bottom of the light socket.

➤ The light socket.

➤ The cord or plug.

The power supply and the receptacle are easy enough to check. A voltage tester will tell you whether the receptacle is hot, and a quick trip to your fuse box or service panel will tell you whether the problem is originating there. If a breaker has tripped or a fuse has gone out, you need to discover the cause before you can consider the problem solved.

The contact tab at the bottom of the socket is the same one referred to previously in this chapter in the "Inspect First" section. Simply unplug the lamp and use a small screwdriver to pull up slightly on the tab. If the bulb still doesn't work, you'll have to look at your cord and plug more closely.

Bugged by Bad Plugs

An intact plug has straight prongs, a solid casing, and a cardboard insulating face-plate or disc (unless the entire plug is solid plastic). If the prongs are bent, the casing cracked, or the faceplate missing, replace the plug.

Bright Idea

It's especially important to re-place damaged plugs and cords on power tools because they draw so much amperage. If you can't do the repair immediately, cut the plug from the end of the cord so no one else inadvertently uses the tool while it's still dam-aged. Do this for lamps and small appliances as well.

Plugs come in a variety of styles including …

➤ Flat-cord plugs.

➤ Quick-connect plugs.

➤ Polarized plugs.

➤ Round-cord plugs.

The easiest plug to install is a quick-connect plug. The prongs of the quick-connect plug can be removed from the casing by squeezing them together. By spreading them, the lamp cord can be inserted. To at-tach a quick-connect plug, squeeze the prongs to-gether and slide them back into the casing. You don't have to do any work on the wire; just follow the in-structions on the package. Be sure the plug is rated for the load to which you're attaching it. (You won't get away with a quick plug on a heavy-duty portable ap-pliance, for example.)

To replace a flat-cord plug, disassemble the casing on the new plug, pull apart the two halves of the lamp cord to the length of two inches or so, strip the insulation off the ends of each half, wrap the wires clockwise around the screw terminals, and reassemble the casing.

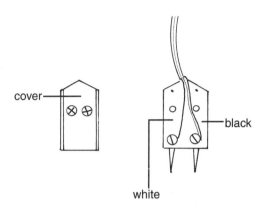

Flat-cord plug.

cover
black
white

The quick-connect plug is the easier of these two to install, but it's not the best choice for a lamp that's unplugged often.

Round-Cord Plugs

These are made for heavier-duty loads and loads needing a grounding plug. The terminal screws also are more substantial than on smaller, flat-cord plugs such as those on lamps and radios.

Follow these steps to install a round-cord plug:

1. Cut the cord end clean with lineman's pliers or a combination tool.

2. If there's an insulating disc with the new plug, remove it.

3. Pass the cord through the clamp on the rear of the plug.

4. Strip two to three inches of insulation from the round cord and ¾-inch of insulation from the hot, neutral, and grounding wires. Tighten the clamp.

Positively Shocking

A lamp cord is divided into two halves. One half carries the hot conductor; the other carries the neutral. The neutral is inside the ribbed or ridged insulation. This is important to know when installing a polarized plug. The neutral wire must be attached to the wider of the two prongs.

5. Take the black wire and the white wire and tie an underwriter's knot as close as you can to the cut edge of the insulation on the cord.

6. Wrap or hook the end of the black wire around the brass terminal screw going in a clockwise direction. Do the same with the neutral wire and the silver terminal as well as the grounding conductor and the green terminal.

7. Tighten the terminal screws securely. Make sure the wires are not touching each other.

8. Install the insulating disc.

Sock It to Your Socket

In addition to damaged plugs, lamp sockets also are a regular culprit when a lamp won't light. The socket itself doesn't wear out, but the switch does. Several types of replacement sockets are available including ...

➤ Pull chain.

➤ Twist knob.

➤ Push lever.

➤ Remote.

It's easiest to replace your existing socket with one of the same style. Follow these steps to inspect and replace your socket:

1. Unplug the lamp.

2. Look at the outer shell of your socket. If it says "Press," you can squeeze the socket and pull it out. If not, your socket is held in by screws, or the socket and harp (the frame for the lampshade) are secured to a threaded metal tube that runs the height of the lamp and is bolted at the base.

3. With the socket removed, check the wire connections at the screw terminals. If they're tight, loosen the screws and remove the wires.

4. Test for continuity by placing your tester's clip on one prong of the plug. Put the probe on the black wire and then the neutral wire. Put the clip on the other prong and repeat the test. If the tester's bulb doesn't light for either prong, you've got a bad cord and plug, and they'll need to be replaced.

5. When you've determined that the socket is the source of your lamp woes, re-place it with a new one with matching volt and amp ratings. Remember to at-tach the neutral wire in the ribbed or ridged insulation to the silver screw terminal and to attach the black or hot wire to the brass screw terminal.

6. Slide the insulating cardboard sleeve and outer shell over the socket and install in the lamp snuggly. (It fits into the lamp cap.)

7. Reinstall the harp, shade, and bulb and then test.

A Dose of Reality

It's great to know how to repair old lamps, but with so many dual-income couples working their brains out all week, you really have to consider whether it's worth your time to do so. Some lighting stores have repair services, and this might be a more eco-nomical way to go once you figure in your time doing the labor and procuring the parts. You might prefer to simply replace the lamp. I'm not advocating consumer waste here, but I understand modern reality. It's one thing to save $40 an hour by pulling your own electrical cable or by installing a needed circuit, but it's quite another to mess around with a lamp socket on a $15 lamp. Hey, how much is *your* time worth?

The Least You Need to Know

➤ Look for obvious problems such as burned-out lamps and loose wire connections before deciding a fixture is broken beyond repair.

➤ Every fixture must have an electrical box, even if the fixture you're replacing doesn't currently have one.

➤ Fluorescent fixtures are reparable.

➤ You might be better off replacing old fluorescent fixtures with new fixtures.

➤ The most common repair for table or floor lamps is replacing the plug, and this job can be made easier by quick-contact plugs.

Chapter 15

Working Around Existing Wiring

In This Chapter

➤ Different types of wiring

➤ What you should replace

➤ Getting inside your walls

➤ Adding a new circuit

➤ Repairing and patching your holes

The easiest residential wiring to work with (short of metal conduit, which you'll never find in a typical home) is grounded NMB copper cable. It's newer than knob-and-tube wiring, the insulation is tough plastic, and you have none of the safety dilemmas that you have with aluminum wiring. If your electrical system was recently installed and inspected, it should be simple to trace circuits and calculate loads as you plan your additions and changes.

Regardless of the age of your system, you'll have to tally your amperage usage for the total system as well as for each individual circuit affected by your work. Simply looking inside your service panel and counting the breakers or looking for an empty space to install a breaker isn't enough. Later in this chapter, I'll discuss how to calculate your additional load and your existing loads and how they affect the electrical service.

Wiring or changing your system generally means doing some damage to your house. The larger the access holes and openings for boxes, the more repairs you'll have to do. Your working goal should be to keep the repairs and patching to a minimum. As you can see, there's more to electrical work than simply deciding to add a receptacle and running some wire. This chapter will take you from start to finish through some small, but important, jobs around your house.

The Dangers of Overextending Yourself

Your electrical system only allows a certain amount of current into your home. A 125-amp service cannot be coaxed into giving you 150 amps. Circuits have their limits, too. In fact, an individual circuit's continuous load should not exceed 80 percent of its capacity. The code is written this way to allow for occasional use of energy-eating appliances or similar loads without tripping a breaker or blowing a fuse.

Your system is divided into branch circuits to distribute the power provided by your utility company. Every load has to be calculated before you can add it to your system. Even if a circuit has not been maxed out and appears to have room for another receptacle, this doesn't mean you can just barge in and add one.

Your Total Load

Electricians do a load calculation, which provides a reasonable estimate of the power to be used when the actual connected load is unknown. The code has the following guidelines for doing these calculations:

1. Multiply the square footage of your home by three watts per square foot.

   Example: 2,500 square feet × 3 watts = 7,500 watts

2. Add 1,500 watts for each small-appliance circuit (minimum of two).

   Example: Three small-appliance, 20-amp circuits in the kitchen total 4,500 watts. One circuit for a workroom is another 1,500 watts.

   Total: 6,000 watts

3. Add one laundry circuit at 1,500 watts.

4. Add the total volt-amps (watts) for items 1 through 3.

   7,500 + 6,000 + 1,500 = 15,000

5. Take the first 3,000 watts at 100 percent, take the next 3,001 to 12,000 watts at 35 percent, and add the two figures.

   Example: We have 15,000 watts so far.

   3,000 at 100 percent = 3,000

   12,000 at 35 percent = 4,200

   Total: 7,200 watts

6. Add the total volt-amps (watts) for fastened-in-place appliances.

   Example:

   Dishwasher: 1,000VA

   Disposer: 800VA

   Electric hot-water heater: 4,000VA

   Total: 5,800

   Note: If you have four or more fastened-in-place appliances, multiply their total VA by .75 and use that figure in your calculations.

7. Add bathroom heat/vent/light units.

   Example: Three units at 1,500 watts each for a total of 4,500 watts.

8. Total the lighting and small-appliance load: 17,500 watts

9. Add the following:

   Electric dryer: 5,000 watts

   HVAC:

   Electric furnace: 13,000VA

   Air conditioner: 6,200VA

10. You can use the larger value—in this case, the furnace—for your calculations because only one system will be run at a time.

11. Add 12,000VA for an electric range. (You also must add on for any additions such as a warming drawer.) The code says a single range can be derated to 8,000 watts.

12. Add on additional equipment such as sump pumps, hot tubs, attic exhaust fans, freezers, and pool and spa equipment.

13. Total up all your loads and divide by 240 volts to get your connected load.

    Example:

    ```
        7,200
        5,800
        4,500
        5,000
       13,000
     +  8,000
       43,500 watts
    ```

    43,500 ÷ 240 = 181 amps

    181 amps is comfortably within a working range of a 200-amp service.

The last step is to make sure the loads in the panel are balanced. That is, you must have roughly the same amperage breakers on each side of the panel (odd-numbered breakers and even-numbered breakers).

Don't Estimate, Calculate!

This is no place for guesswork. Remember the following formula to figure out the appropriate load for a circuit:

watts ÷ volts = amps

This means totaling all the loads or anticipated loads on a circuit (light bulbs, blenders, TVs, tanning beds, and so on) by their wattage and dividing the total by the voltage of the circuit (either 120 for most house circuits or 240 for heavy appliances). The total, in amps, will dictate the required rating of the circuit (15 amps, 20 amps, 30 amps, and so on).

The wattage demand on the circuit should equal 80 percent of its maximum capacity. A 20-amp, 120-volt circuit can provide a maximum of 2,400 watts (volts × amps = watts). Eighty percent of 2,400 equals 1,920 watts, which sometimes is rounded off to a working load of 1,900 watts. This is why it's important to map out your circuits so you know what they're running and can determine whether you can add on to them (or whether you should remove a load).

Electrical Elaboration

In our last house, a 1924 bungalow, we discovered that the washing machine was running off the basement light circuit. We traced knob-and-tube wiring tied into NMB cable and a short section of it running in an old pipe used as conduit. Needless to say, we rewired it all.

Leave Dedicated Circuits Alone

Don't mess with a dedicated circuit. You might be tempted to add a living room receptacle on to your small-appliance circuit in the kitchen because it's convenient, but this is where the trouble starts. It's against the code to use these circuits for anything except their designated small appliances. Do this now and someone will have to undo your work later.

Uh-Oh, Old Wiring

You're down in your basement looking at a rat's nest of electrical wiring, junction boxes, and questionable splices. Your house has a new service panel, but few if any circuits were replaced. How much of your old wiring can you save, and what should you gut out?

Before you start snipping away at every errant wire in sight, step back and take a realistic look at your system. Pull out your circuit map and trace the worst-looking wiring to a device or an appliance so you know what's running off of it. In some cases, it's completely acceptable to run new cable from the service panel and tie it into existing knob-and-tube wiring inside a junction box. (Be sure the knob-and-tube wiring is covered with its original loom insulation.) In other cases, you can pull the cable to the first device on the circuit (a receptacle, for example) and keep the remaining wire intact.

Remember, the main problems with old wiring are bad or weak connections and misguided alterations by homeowners or contractors. By undoing this work and making sure remaining connections are tight and secure, you can enhance your existing system and ensure its safety.

Mixing Old and New Wiring

You can connect new cable or wire to existing cable or wire if you follow these rules:

➤ All connections must be made inside an electrical box. The one exception is knob-and-tube wiring, which is the only electrical system that can be spliced in the wall. To do so, you must solder the conductors or use a mechanical splice such as a split bolt (see Article 324 of the NEC); a wire nut is not sufficient.

➤ Junction boxes must be kept accessible. They cannot be covered up.

➤ Don't change the wire gauge. (New wire has to match existing wire.)

➤ Be careful not to overextend the circuit by adding more loads than it's designed to handle.

Positively Shocking

When connecting a line-side NMB cable with a grounding conductor to knob-and-tube wiring inside a metal junction box, you must ground the box. If the NMB is on the load side, leave the grounding conductor alone. In a plastic box, you can leave the grounding conductor loose. Never attach an ungrounded grounding conductor to any ground screw. This implies a circuit is grounded when it isn't.

Bright Idea

When running new cable from a panel to a junction box to connect with existing wire, be sure to mark the inside lid of the junction box with the breaker number. You want to leave clear instructions for future electricians.

Knob-and-tube wiring doesn't lend itself to easy identification of the hot and neutral conductors. You might have a junction box or device box packed with wires. Identify the hot lead the same way you would identify it in a less-crowded box:

1. Identify the circuit and turn off the power at the panel.

2. Check the connections with a voltage tester to confirm that the power is off.

3. Carefully remove any tape or wire nuts from the connected wires (those running back toward the panel).

4. Mark the wires so you know which ones were connected to each other.

5. Separate the wire ends so they're not touching each other or the sides of the (metal) box.

6. Turn the power on at the service panel and test the wires one at a time until the hot lead lights the bulb on the tester. Mark this as the hot line conductor.

Messy junction boxes often indicate that a device or fixture was added without a lot of consideration as to its effect on the circuit. Your best bet is to confirm whether the circuit can or cannot safely support the addition and then deal with it appropriately. It might be that all you're looking at is an unkempt, but safe, series of connections.

Positively Shocking

Aluminum wiring presents special considerations. You should have a qualified electrician inspect and modify any aluminum wiring.

How Much You Can Keep?

Basically, you can retain any and all safe wiring and devices. An electrical system will do what it's designed to do and will do so reliably until you extend a circuit or the system beyond its design. You *should* remove or replace the following:

➤ Any corroded or damaged wiring

➤ Wiring that has not been installed properly (such as wire running along the bottom edge of an exposed joist rather than through holes drilled in the joist)

➤ Devices or fixtures that render a circuit unsafe

➤ Conductors that are the wrong gauge for their circuits

➤ Any cable rated for interior use but installed outdoors

You might be getting more than you bargained for if your system has been repeatedly altered over the years. If additions have been made based on convenience ("Hey, look! I found a couple of wires.") rather than logic, you could end up with an odd variety of lights, receptacles, and other fixtures all over your house. It might be best to run a new circuit or two to cover all these miscellaneous runs if you're not already rewiring the whole house.

Walls Everywhere

Running cable through finished walls or ceilings is a nuisance. Once you've determined that a circuit can be extended or a new one added, you should plan the circuit's route and figure out the least-disruptive route for the cable.

An existing circuit can be extended from an electrical box provided that ...

➤ The box is sized to accommodate the additional cable.

➤ The box isn't at the end of a switch loop (an end-of-the-run switch).

➤ The box isn't a switch-controlled receptacle (unless you want the added device to be controlled by the switch as well).

Your best and easiest route is through an unfinished basement or attic, as shown in the following figure.

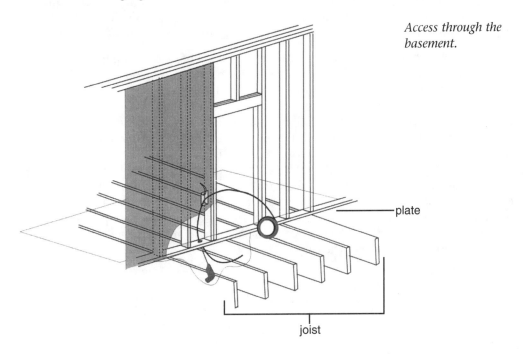

Access through the basement.

plate

joist

Going fishing: Use the fish tape to pull the cable through studs.

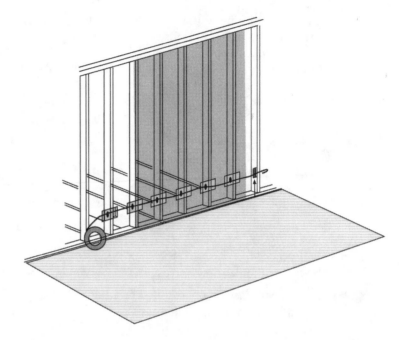

If you're drilling into a crawl space, place a flashlight over the top of your drilled hole and turn it on. This makes it a lot easier to find the hole when you're crawling under the house. If you have several holes to drill, insert a piece of scrap neutral wire into the hole; you'll be able to see its white insulation more easily and mark each hole.

It isn't always easy to determine the location of a box when you're up in an attic or down in a basement. From an approximate location in an attic, you can drill a very small hole on the outer edge of the wall's top plate through the ceiling below and then poke a section of wire hanger or scrap wire through. From the room below, you can locate the necessary drilling location, go back to the attic, and adjust your coordinates. From the basement, you can drill a pilot hole from above (if you're adding a receptacle) through the sole plate or bottom plate of the wall using a long, narrow twist bit. Then drill the larger hole from below. If the floor is only covered with subflooring material such as plywood, you can use your drill or drive a 16d finish nail through a location near the base of the wall where your new box is going in, looking for the nail down below. If you don't object, you can do the same through carpet; just be sure to place the nail near the very edge of the baseboard and use a *nail set* to pound the head below the surface of the carpet.

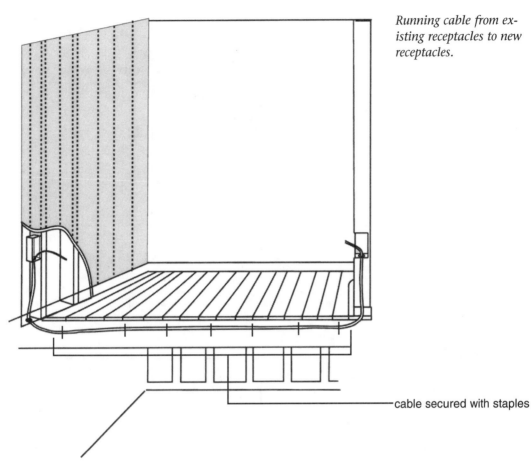

Running cable from existing receptacles to new receptacles.

cable secured with staples

What if you have wood flooring? If you cannot accurately determine where to drill from the basement, drill the smallest hole possible (with a drill bit about four inches long) near the edge of the baseboard. Push a piece of wire through the hole so you can find the location in the basement. A piece of scrap conductor from your NMB cable will work just fine. At the end of the job, fill in the hole with a putty stick in a color matching the floor stain.

Ask an Electrician

A **nail set** is a steel punch used with a hammer to set the head of a nail below a wood surface. With the head set, the nail hole can be filled with putty or other material to hide the nail. This typically is done as part of finish carpentry.

Using a small sash chain, pull the cable up and through the attic joist.

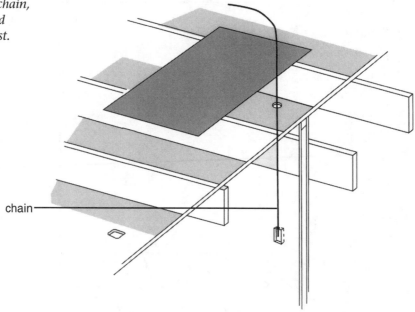

chain

Drill or nail close to the baseboard.

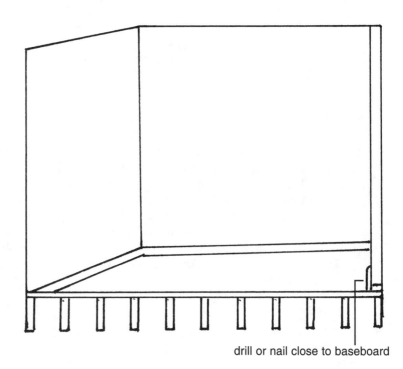

drill or nail close to baseboard

Boxes and the Code

The code wants to control the spread of fire. On fire-rated walls (between the garage and any living areas), electrical boxes must be mounted at least 24 inches apart (unless a fire rating is maintained between the boxes, which is unlikely in residential construction). Among other things, this means you cannot mount boxes back-to-back. Boxes must be used for any wire splices or connections, and they must be sized to comfortably house the number of wires making those connections. All boxes need cover plates.

The Perils of Plaster

Plastering is quite a craft. Older plaster jobs consist of three coats of material applied over wood or metal lath. Some old plaster (Victorian homes come to mind) is a little on the crumbly side. If you're careful cutting through it, you'll keep your repairs to a minimum.

To cut through plaster ...

1. Determine the location of your new box (next to a stud if possible, but do your initial cutting at least a few inches away to make sure nothing gets in the way of the saw blade).
2. Place a drop cloth or a piece of plastic on the floor.
3. Drill a test hole so you can determine where the edge of the stud is located. (Ignore this if you're not going to be near a stud.)
4. Place the front of your new box against the plaster, and use a pencil to draw around it (ignoring any plaster ears) to give you a line to cut into. You can apply masking tape around the outline to help keep the plaster from chipping when you cut.
5. Drill a hole in each corner to provide starter locations for the keyhole saw or scroll blade.
6. Hold a straightedge against the pencil lines, and score the plaster several times with a sharp blade.
7. Carefully saw through the lath in smooth movements. Go through about $\frac{7}{8}$ of the way on one side and then cut the other side completely, returning to cut the remainder of the first side. This prevents the lath from excessively shaking the plaster.

If you have metal lath, you can't really saw through it. You can try to chisel it out, but this can cause further cracking if you're not careful. I'd just drill as many holes as you need, following the pencil outline to minimize the chiseling. Once the metal lath is exposed, cut through it with tin snips and smooth out the plaster with a saw blade or a rough file, or tap it with a small hammer.

Cutting plaster and lath with a keyhole saw.

tape ———————————————————————

keyhole saw ———————————————————

drill holes in each corner   score lines

Drywall

Cutting through drywall is a lot easier than plaster because there's no lath to deal with and the material itself is softer than plaster. After outlining the shape of your electrical box, you can cut through drywall with a keyhole saw, a small drywall saw, or a hacksaw blade. You also can cut through it with a strong, sharp knife using sawing motions if you don't have a saw available.

Positively Shocking

Be careful when working in an unfinished attic. If it's an open floor, you'll need to place a board or a piece of plywood across the joist to support you. Wear an appropriate mask or a respirator if you have to work around insulation.

Boxes in Ceilings

Light fixtures and fans require an electrical box installed in the ceiling. (Fans are heavy and require special boxes. If you use anything else, the fan might fall out of the ceiling.) An unfinished attic gives you plenty of access to the joist. All you have to do is drill a small pilot hole at the proposed box location into the ceiling from below and poke a section of metal hanger up and into the attic (check the approximate location in the attic first for obstructions and wires). This will enable you to see if there are any obstacles in the attic such as existing wiring or framing that might be in the way of a box. Once you've established that the location will work, you can either …

➤ Cut the round hole for the box with a hole saw.

➤ Use a keyhole saw or a drywall saw for cutting the hole.

Ceiling boxes can be either nailed to a joist or attached with bar hangers. Nailing to a joist is simpler, but your light might fall between two joists and thus require hangers.

No Access, Now What?

Finished basements and attic spaces (or service panels located in finished garages) call for a more deft approach to your wiring. You have to decide if you should …

➤ Run part of the circuit in conduit across the basement or garage ceiling.

➤ Run conduit on the outside of your house and then into the walls.

➤ Tear into the walls and ceiling at regular intervals, exposing the studs and joist to run the cable.

Most studs and joist are spaced at 16 inches on center (which means the center of the nailing side of one will be 16 inches from the center of another). The spacing offers predictable nailing surfaces for drywall and interior trim, especially baseboards. If you have no other way of getting into the wall or ceiling, you'll have to open up the wall on each side of the studs and joist until you've reached the locations for your new boxes.

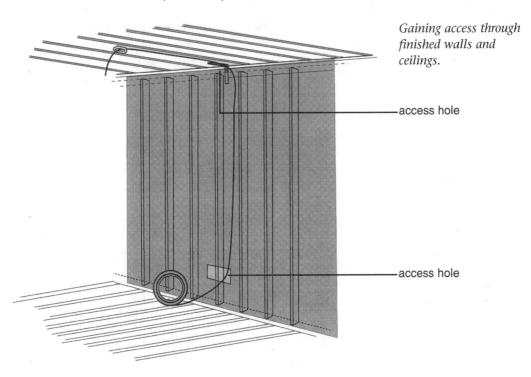

Gaining access through finished walls and ceilings.

access hole

access hole

The Woodwork Comes Off

Cutting very visible holes in your walls and ceilings (and repairing them later) might not be at the top of your list of fun projects. If you have wide baseboards, you might have a way out. You have to decide if it's less of a mess to carefully remove a length of baseboard, pull the nails, reinstall, and possibly repaint it. With the baseboard out, you can drill behind it and not bother with patching up the plaster or drywall. This only works with wide baseboards (six inches wide and wider). With narrow baseboards, you would have to drill too close to the bottom of a stud, which could weaken it.

Electrical Elaboration

Old woodwork can crack and split around nails if it isn't carefully and slowly removed. Gradually work it away from the wall using a thin pry bar or flat bar and a hammer. You can even take a hacksaw blade and cut the nails when the board has been backed away enough to get at them. If you use a hammer for part of the removal, place a board against the plaster when you pry; otherwise, the hammer will damage the wall.

Bright Idea

Drilling leaves sawdust and shavings at the bottom of your wall cavities. If possible, vacuum these out. There might be only a remote chance of a device shooting any sparks off in the future, but you don't want them falling on easy-to-burn wood waste.

Get Your Drill Out

The smaller the drill, the tougher the job when drilling holes for running cable through your house's framing. I've already discussed drills (see Chapter 8, "Call Me Sparky"). When access is limited, a right-angle drill can be very advantageous. *Your holes must be drilled at least 1½ inches from the edge of the stud or joist!* This is to avoid damage to the cable from dry-wall nails or screws or any other fastener used to attach anything to a wall or ceiling. If the studs are exposed, you should install a metal nailing plate in front of the hole for added protection.

Insulation Obstacles

Insulation is great for maintaining your house's temperature, but it's terrible to pull wires through, especially the blow-in cellulose type. Worse yet is solid foam.

You'll never get a cable through that stuff. If you have insulation and you're not inclined to pull your walls apart and remove it, consider …

➤ Using some form of raceway or conduit.

➤ Installing floor receptacles.

➤ Looking for alternative locations on interior walls instead of exterior walls.

➤ Going through closets or cabinets.

The Finish Work

Electrical boxes must be set so they're flush with the surrounding wall or ceiling when wood framing is used for construction (NEC Section 370-21). The gaps around the box cannot be greater than ⅛ inch; anything greater must be filled with patching plaster. The code wants to diminish the possibility of a fire spreading if a device or fixture has a short circuit. Boxes are secured by …

➤ Nailing to the stud or joist.

➤ Plaster ears.

➤ Snap-in brackets.

Metal boxes can be used with …

➤ Nonmetallic cable.

➤ Armored cable.

➤ Metal conduit.

Nail-on plastic or nonmetallic boxes can be used with nonmetallic cable. Others are rated for use with plastic conduit.

Positively Shocking

Use the right type of box for the job at hand. Surface-mounted ceiling lights require a round or octagonal box. Recessed lights often come with their own junction box, while fluorescent fixtures need their attached wires to be connected with the circuit cable in a separate junction box.

Fitting All Those Wires In

There's no point in using the minimum-size box necessary for the job when a larger one will make your work easier. If it's an aesthetic issue, which is doubtful, by all means go with the smaller box. In either case, it must be the correct size for the job (see Chapter 11, "Switches and Receptacles"). The code doesn't like to see too many wires being stuffed into too small of an electrical box. A larger box also gives you some leeway in the future should you need to run additional wires into it.

Plaster and Drywall Repair

Plaster is applied wet and is forced into wood or metal lath. As it dries, the material that has oozed onto the back side of the lath forms plaster keys that hold the plaster in place. After you've cut the lath away to run your cable, you have to either patch in new lath for backing or use a small section of drywall for your repair.

The cleaner your access holes, the easier your repairs. You might have to repaint the entire wall (maybe the rest of the room, too), depending on how much the paint has faded.

Patching a hole in drywall is a little easier than doing the same with plaster. All you need to do is …

➤ Even up the hole in the wall by tracing an outline an inch or so in each direction. Take these dimensions and cut out an even rectangle or square of new drywall.

➤ Hold the drywall patch against the wall, trace around it, and cut out an even, level hole.

➤ Cut a second drywall patch two inches larger in each dimension.

➤ Drill a hole in the center of the larger patch and pass a cord through it, knotting one end so it holds tight against the hole.

➤ Apply construction adhesive to the outer two inches of the face of the patch.

➤ Insert the patch into the wall and align it with the hole. Pull it tight to the wall with the cord.

➤ Tie the other end of the string to a pencil or a small piece of scrap wood, and twist it until the patch holds tight. Tape the end of the wood to the patch so it doesn't slip. (You also can screw through the wall and secure the patch.)

➤ After the adhesive has set, test the smaller patch for fit and trim down the edges if necessary. Drill a hole in the middle of the patch.

➤ Pull the cord through the second patch, and glue the smaller patch to the larger patch, securing it by twisting the cord.

➤ After the patch has set, cut the string and push any excess back into the hole. Apply fiberglass drywall tape to the edges, and cover the patch with a thin application of joint compound.

➤ When it's dry, lightly sand the patch with 100-grit paper and apply a second coat of joint compound, spreading it slightly beyond the tape. Sand it when dry and apply a third coat if necessary.

➤ Run your hand over the patch after each layer of joint compound has dried. (You don't want to end up with a lump.)

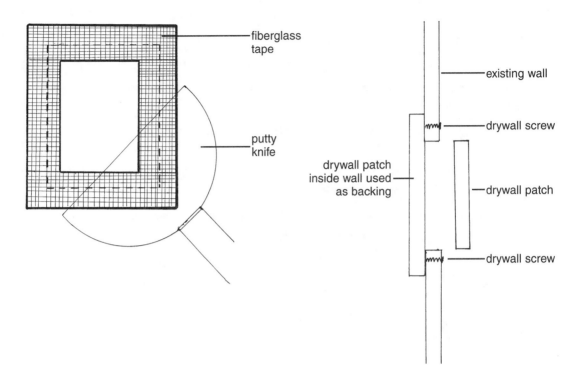

fiberglass tape

putty knife

existing wall

drywall screw

drywall patch inside wall used as backing

drywall patch

drywall screw

Wall patches.

Because of the plaster keys, you won't be able to easily glue a drywall patch to the back of the lath inside the wall cavity. I suggest putting extra construction adhesive on the patch, pulling it tight, and securing it with screws through the plaster. Be sure to countersink the screws so you can conceal the heads with joint compound.

Bright Idea

The easiest way to remove substantial loads from existing wiring is to install dedicated circuits for the refrigerator, laundry, and kitchen circuits.

The Least You Need to Know

➤ Always calculate the existing and potential load of a circuit before you add on to it.

➤ Old wiring isn't automatically evil. You can remove the worst of it and safely incorporate the rest with new cable.

➤ The more carefully you plan your work and routing for circuits around finished walls and ceilings, the fewer repairs you'll have to do later.

➤ Be certain that any wire, device, or fixture you add to an existing circuit is rated for the ampacity of that circuit.

➤ Allow for extra time at the end of the electrical job to repair and repaint dry-wall or plaster.

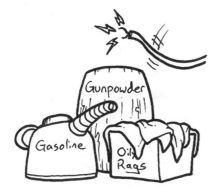

Chapter 16

Trouble, Troubleshooting, and Safety

In This Chapter

➤ Clues to electrical problems

➤ Paying attention to the warnings

➤ Playing it safe

➤ Precautions when the power is out

Residential electrical systems, especially newer installations or upgrades, usually just keep humming along. Cartoonists could depict billions of smiling, happy electrons zipping around our wires, one little electron holding hands with the next, doing our electrical bidding. But things can go bad. Bare wires can cross each other, appliances can short out, and lamp cords can become frayed.

Your electrical system, unlike the institution of democracy, might not require constant vigilance, but you have to keep an eye on it. Circuit breakers that trip regularly and fuses that burn out too often are signs of a problem circuit. Dimming lights are a romantic touch when you control them with a dimmer switch but not when they dim on their own. Likewise, if your electric can opener shoots sparks like a Roman candle, it means you have a problem.

This chapter will serve as a review and will cover some of the basic warnings and problem-solving skills you'll need when you run into electrical mishaps. Safety will be emphasized as always, and you should pass this emphasis on to your children and anyone else living in your house. Thousands of people are electrocuted every year,

Bright Idea

Sometimes the problem *is con- tact.* Make sure the terminal screws are tight and making full contact with their conductor. In lamps and lights, make sure the bulbs are making complete con- tact with the sockets. Don't go crazy diagnosing your system when a simple repair might be all you need.

Positively Shocking

If an appliance or another mo- torized load suddenly stops working, especially after you hear a popping sound, do not attempt to restart it. Any burning smell will tell you that the motor has shorted or is otherwise dam- aged. Turn off the circuit and then pull out the plug before doing any repairs.

many of them children, and they often require hos- pital care. A safe, monitored electrical system will pre- vent many needless injuries and possibly even deaths. Without any further lecturing, we'll move on to scru- tinize your wiring, devices, and appliances.

The Rules Once More

These are the basic safety rules you must remember:

➤ Never work on a live circuit, fixture, or device.

➤ Shut off power to a circuit before repairing a de- vice or load on the circuit. Keep one arm behind your back (or otherwise away from the panel) when shutting off or turning on a circuit.

➤ Use a fuse puller to remove and replace fuses. Consider shutting off the power to the fuse box at the main disconnect.

➤ Tape over or otherwise mark a main switch, fuse socket, or circuit breaker that's been shut off if you're working anywhere on your electrical system. This tells others not to turn the power on. (Post a large note on the fuse box or service panel as well.)

➤ Always test to make sure the circuit has been shut off before doing any repairs.

➤ Always unplug a cord-and-plug appliance, lamp, or other similarly connected load before repairing it.

➤ Never stand in a puddle or on a wet surface when doing electrical repairs. Place a piece of wood on a damp floor and wear thick rubber boots to insu- late your feet.

Warning Signs

Life is full of warning signs, but we don't always pick up on them. A circuit that constantly burns out its fuse does not need a larger-amperage fuse. Instead, reduce some of the loads and their demand for current. Remember, electrical systems are very logical and safe when used intelligently. If you push them beyond their limits, all bets are off as to how well they'll behave.

An orderly electrical system doesn't overheat, start fires, or inordinately dim your lights. These are all signs that you have problems. (A fire means you're too late.) If you have an older system that still uses fuses and does not have a grounding conductor, you need to be more observant of your usage and how your system reacts.

Electrical problems sometimes require detective work. When an appliance doesn't work at one receptacle, try it in another before you start tearing it apart. It might just be a bad receptacle or a loose terminal screw.

Remember that older, fuse-based systems weren't designed for all the electrical loads we surround ourselves with today. These systems are safe when used judiciously.

Hot Stuff

The cover plate on a light switch or receptacle should not feel excessively warm and certainly should not be hot. A dimmer is an exception because dimmers dissipate the heat from dimming through the fins of the dimmer and often transfer some of that heat to the screws holding on the cover plate. Cords and plugs shouldn't feel hot, either. Heat is a sign that the load is demanding current in excess of the ampacity of the electrical cable and/or the plug and cord attached to the load. If you replace a 60-watt light bulb with one of a larger wattage (one whose wattage exceeds the rating for the fixture), the wire or cable will still supply the current, even if doing so makes the fixture dangerous.

Heat signals that you should examine the total load on a circuit or a cord and plug. In the case of a hot circuit, you should ...

➤ Make sure the circuit breaker or fuse is the correct amperage for the circuit itself and the cable or wire that forms the circuit.

➤ Total the combined load on the circuit.

➤ If the load exceeds the circuit's design, reduce the load.

The real danger of an *overloaded circuit* is a wire heating up unseen inside your walls to the point where it can start a fire.

Ask an Electrician

An **overloaded circuit** has more load(s) and demand for current than it is designed to handle, creating a potential fire hazard. A **short circuit** occurs when a bare hot wire touches a bare neutral wire, a bare grounding conductor, or another piece of grounded metal such as the side of a metal box. The flow of extra current should trip a breaker or blow out a fuse.

Bright Idea

Some appliances cause a momentary power surge when their motors first start up. This surge can blow out a regular fuse. Consider installing a time-lag fuse that allows for these brief overloads and waits a second or so without blowing.

Short Circuits

One telltale sign of a *short circuit* is black, smoky residue on switch or receptacle cover plates. Frayed or damaged cords and plugs also can be sources of short circuits. You need to check further for the source of the problem if your circuit goes dead and ...

➤ You cannot find any visible signs of an electrical short.

➤ The circuit is not overloaded.

Before replacing the fuse or resetting the breaker, turn off all the loads and unplug everything from the receptacles. If the new fuse blows or the breaker trips right away, your problem is either in one of the devices (a switch or a receptacle) or in the wiring itself.

To make sure the current is dead, remove all the cover plates and examine each device for charred wires or black residue. Clip the ends of any affected wires, strip off sufficient insulation, and install a new device. Replace the fuse or reset the breaker, and test the circuit again.

What if the circuit doesn't short immediately after you set the breaker or replace the fuse? In that case, activate each load one at a time and then turn each one off. Check the load that eventually causes the short. The problem will be either in the fixture or appliance itself or in its wiring. Replace the offender and check the circuit again. If it still shorts out, you have a problem in the wiring itself and should call an electrician. Shorts in the wire almost always are at the device or fixture box, so the problem should be visible when you do your own inspection. Sometimes, however, the problem is caused by a splice or junction box buried in the wall and is therefore unnoticed upon first inspection.

In the case of a plug-in appliance or lamp, if the circuit goes dead as soon as you insert the plug, you can assume the short is in the cord or the plug (both of which can easily be replaced). If the short doesn't occur until after the appliance is turned on, the problem isn't in the cord or the plug but in the appliance itself. You should then repair or replace the appliance.

Power Cords

Any cord-and-plug combination is subject to damage and wear. Lots of pulling and twisting, especially when a plug is pulled out by the cord rather than by grasping the plug itself, can cause the wires inside the insulation to break. This is particularly true with lamp cords because they use wire composed of multiple strands with a very small gauge. If the insulation protecting a cord cracks open, you could get a short circuit. It's usually a better idea to replace these cords than to repair them (see Chapter 14, "Light Fixes").

Although it is less likely that you'll ever have to replace a 240-volt appliance cord, these also can be changed out:

1. Unplug the cord and unscrew the end that's attached to the appliance, noting which wire (by color) went with which screw.

2. Buy an exact replacement for the cord and plug.

3. Connect the new cord to the terminal screws on the appliance, noting any color coding on the screws to match the individual wires (black to black, and so on).

Positively Shocking

Remember that aluminum wiring requires the use of devices marked CO/ALR only. Use CO/ALR devices to replace older, unmarked devices or those marked CU/AL. Never back wire devices with aluminum wiring. You should only use the terminal screws and only after applying a deoxidizing agent to the ends of the wires.

Lamps

Most problems with lamps occur in the cords, but occasionally a socket or switch goes bad and needs replacement (see Chapter 14). Always test the light bulb first. If the bulb works after being placed in another fixture, test the receptacle. A damaged table or floor lamp should be removed from the room until it's repaired. A light fixture should be tagged as broken until it's repaired, and the bulb should be left in place.

Always make sure the receptacle is not switch-controlled. In new homes, to meet the code requirement that calls for one switch-controlled light in every room, builders often install switch-controlled receptacles instead of ceiling or wall fixtures in living rooms and bedrooms. If your lamp isn't working, perhaps the switch is in the "Off" position.

Incandescent Light Fixtures

When a hard-wired light fixture fails to light, check the ...

➤ Power supply.

➤ Bulb.

227

➤ Switch.

➤ Fixture itself.

Always replace devices and fixture components with items that have the same ratings for voltage and wattage.

Electrical Elaboration

Replacing an old light fixture might be a bigger job than you expected if the fixture does not have an electrical box. All fixtures and devices must have a box to contain their wire connections and as a base for securing the fixture or device. This is a safety issue and must be heeded. You might want to call an electrician.

Fluorescent Woes

Fluorescent fixtures have a few additional components to check when they stop working or the bulbs dim. In addition to inspecting the switch and confirming that the circuit is working, check for …

➤ Loose bulb(s).

➤ A bulb that needs replacement.

➤ A defective starter, ballast, or socket(s).

➤ Erratic starting in older homes where the systems do not have a ground.

Fire Hazards

Fuses and circuit breakers are designed to protect you from inadvertently overloading a circuit. When your loads demand more current than the circuit is designed to handle, the circuit breaker will trip or the fuse will blow. This prevents the conductors or wire from overheating and causing a fire. A conductor can only offer a certain amount of resistance to a current; if there's too much current, the conductor can heat up enough to melt its insulation.

A homeowner can create a dangerous situation by replacing a fuse or a circuit breaker with one of larger amperage, thus allowing more current to flow through the wires than they can safely resist. A fire can start without tripping the breaker or blowing the fuse because the larger-amperage fuse cannot sense the problem. Some signs of a potentially overloaded system include …

➤ Thirty-amp fuses used for lighting circuits.

➤ The use of extension cords as permanent wiring.

➤ Dimming lights when appliance loads go on.

➤ Excessive use of adapters that allow more than two loads to be plugged into one receptacle.

➤ Multiple service panels and sloppy wiring practices.

Bright Idea

Are you uncertain about the cause of an electrical problem? Fuse systems can give you a hint. Screw-in plug fuses will tell you why they burn out. An overloaded circuit will melt the metal tab inside the fuse. A short circuit will leave a black residue on the glass. These are important and helpful diagnostic clues.

Don't Overextend with Extension Cords

Extension cords are designed for temporary usage, not permanent or semipermanent installations. The cord you use should be rated for the load it's going to power. An 18-gauge cord cannot safely carry the current necessary to run a portable kitchen appliance, and it most likely will heat up and create a hazardous condition. In addition, you should not …

➤ Use an indoor-rated cord outdoors.

➤ Run an extension cord under a rug or carpet.

➤ Staple or tack an extension cord to anything.

➤ String cords together without calculating the loads.

➤ Use a cord while it's still coiled. (It should be unrolled for use.)

Regular Tests You Should Do

All GFCI receptacles and breakers should be tested once a month. In addition, you should check your smoke detectors at least once a month, especially battery-operated detectors. The batteries themselves should be replaced once a year.

Electrical Elaboration

The most dangerous month for electrical fires is December. Why? Because the holidays bring about increased use of decorative lighting, heaters, and appliances. Visiting relatives add to these increases and to the risk of inappropriate use of electrical devices. It's important that all light cords be checked. Keep flammable materials away from lights, candles, and portable heaters.

General Precautions

Einstein considered common sense to be all the prejudices you acquire before the age of 18. (This was the belief of someone who apparently had to be reminded by his housekeeper to dress warmly before venturing out into Princeton winters.) Semantics aside, a certain amount of common sense should be applied to your electrical dealings.

Precautions for electrical appliances, devices, and wiring include the following:

➤ Buy only items that are UL-listed or approved by another accepted testing agency.

➤ Keep children from playing near portable heaters and kitchen appliances.

➤ Use kitchen and bathroom appliances on or near dry surfaces only.

➤ Keep combustible materials such as clothes and curtains away from heaters of any kind.

➤ Never cut off the grounding pin from a three-pronged plug.

➤ Never file down the larger prong on a polarized plug.

➤ Use child-resistant caps in unused receptacles.

➤ Install smoke detectors.

➤ Make sure the contact between a plug and a receptacle is solid and tight.

➤ Allow plenty of free space around computers, televisions, and stereo sound systems to prevent them from overheating.

➤ Keep metal ladders away from all power lines.

➤ Stay away from any downed power lines.

➤ Have your electrical system inspected if it's more than 40 years old and you have no record of a recent inspection.

➤ Make sure all switches and receptacles have cover plates.

➤ At the very least, install plug-in GFCI receptacles into existing bathroom and kitchen receptacles that do not have grounding.

➤ Make sure light bulbs are the correct rating for their lamp or fixture. (Excessive heat can be a fire hazard.)

➤ Unplug portable appliances when they're not in use, especially those near sinks. (You can be electrocuted if they fall into water—even if they're turned off.)

➤ Leave electric blankets untucked.

Power Outages

You can't do much to control power outages, but you can control what happens when the power comes back on. Unplug your computers and television sets, even if you have surge suppressors. The suppressors should take care of any initial charge from your electrical system, but unplugging these appliances guarantees that you won't have any problems. Make sure any kitchen appliances that might have been left on, other than your refrigerator, are shut off. Any heat-producing appliance, such as an electric blanket, a heating pad, or a portable heater, also should be shut off, lest you forget about it and it stays on while you're not home.

Positively Shocking

If it's incorrectly wired, a GFCI can allow a current to pass through it without providing shock protection. Test the receptacle by inserting a small night-light and then pressing the "Test" button. The light should go out. If the "Reset" button pops out but the light is still on, you need to inspect the receptacle's wiring. You can be electrocuted by an incorrectly wired GFCI.

Bright Idea

Dimango (1-888-766-0333 or www.dimango.com) makes a Power Failure Light, which is a night-light that can be removed and used as a battery-powered flashlight during power failures. AliTer Industries (1-800-505-3161) sells the PowerAlert alarm, which monitors power loss and sounds an alarm for any device plugged into it.

The Least You Need to Know

➤ The number-one safety rule when working around electricity is to turn the power off before working on a circuit. Always verify at the device that the power is off.

➤ Monitoring your system from time to time and reacting to warning signs will keep your system—and you—safe.

➤ Damaged cords and devices should be repaired, replaced, or removed from use.

➤ Be very careful with extension cords. They should not be used as permanent wiring. Also, extension cords enable you to dangerously overload circuits.

➤ If you treat your system in the manner for which it was designed, you should never have a problem with it, let alone a dangerous one.

Part 4

Power Hungry

The earliest homeowners who benefited from electrification might have had a whopping 30 amps of power serving them through a small, wooden fuse box. This was enough to run ceiling lights and receptacles for lamps. Today, we are the recipients of a vast power generation and delivery infrastructure, and we should take every advantage of it. On a residential level, this means a service panel (200-amp when called for) with circuit breakers and plenty of branch circuits.

A new or updated panel enables you to bring safe, grounded circuits to every corner of your house. This is especially important in kitchens and bathrooms and outdoors. Other areas also will benefit from additional receptacles for all the electronic equipment we can't seem to live without. If you still have an old fuse system, the change and the convenience of a new panel and circuits will be noticed right away. No longer will you need extension cords or power strips scattered around the house.

I don't recommend adding new service on your own. This really is a good time to call an electrician. New circuits are another matter, depending on your comfort level with the work. You can always do a portion of the work and then have an electrician finish up. As always, check with your local building department to confirm whether a homeowner is allowed to do this work at all. Some municipalities mandate that certain jobs must be done by a licensed electrician.

Chapter 17

Service with an Attitude

In This Chapter

➤ Installing a new electrical service

➤ Sizing your service

➤ Transition from new to old

➤ Locating the panel

➤ How much to budget

I'm a big believer in updating the basic mechanical and structural functions of a house (electrical, plumbing, heating/air conditioning, foundation, and roof) before just about anything else. It doesn't do much good to have exquisitely painted, papered, and refinished living areas if you don't have enough lighting to see any of it. Electrical standards and minimum code requirements have improved over the years but maybe not in your house.

Do you still have fuses? A small service? Do you dream of being able to turn on your toaster and the coffeemaker at the same time without blowing a fuse? You need a new service. It won't be as much fun to look at or show off as refinishing your hardwood floors, but you and your family will have the start of a convenient, up-to-date, safe electrical system. Most electrical improvements in an old house begin with a new service. (Note: If the existing fuse or breaker panel has room for additional circuits, a new service might not necessarily be needed.)

When you decide to do an upgrade, you'll have a number of issues to consider:

➤ The size, brand, and price of the new service panel

➤ Whether to do some or all of the project yourself or hire an electrician

➤ The location of the new service panel

There are quite a few rules to follow when installing a new service, from how to support the cables to different means of establishing a grounding electrode. Although I think this is best left to an electrician, we'll go through the steps so you'll know what's involved.

One New Service Coming Up

Your service panel is the primary distribution center for all the electrical currents in your house and yard. It connects your CD player, lava lamp, and 27-speed blender to your utility's service lines. Fuses serve the same purpose as the circuit breakers in a modern service panel, but they are considered dated and no longer are installed for residential purposes. Some old fuse services are as small as 60 amps, which is hardly adequate for modern electrical demands. The National Electrical Code calls for a new service to be a minimum size of 100 amps.

There are two types of electrical services:

➤ Overhead service

➤ Underground service

Most older services are overhead. That is, the utility company uses overhead service conductors, usually from a utility pole, that often connect to a service mast on the outside of your house. An underground service, commonly found in new housing, is buried. Each must follow prescribed installation procedures.

Overhead and Exposed

It's a lot simpler to install an overhead service in an established neighborhood than it is to start tunneling under streets and sidewalks to run conductors underground. Overhead wires are exposed to the weather, however, and this means your service can be disrupted if a tree branch falls on it during a high wind.

Modern electrical service consists of two hot conductors and one neutral conductor coming into your home. They come out of a transformer, which steps down the voltage, and must clear roofs, fences, and outside structures as they connect to your service head on top of your service mast. The conductors form a *drip loop* as they enter the service head so that any rain landing on them will not run down the mast.

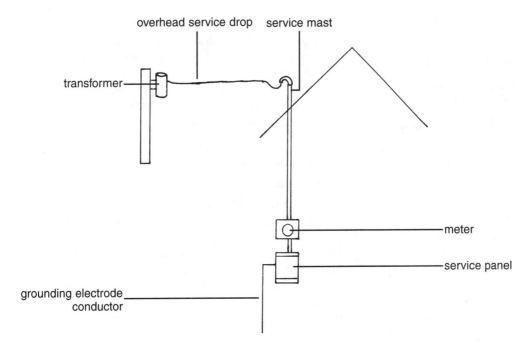

Overhead drop service.

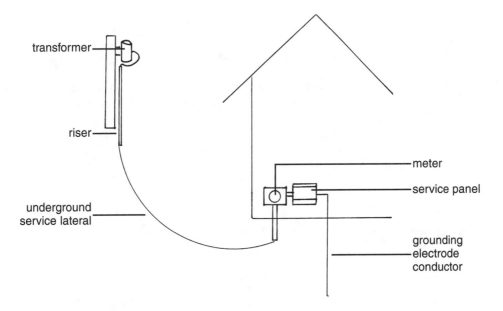

Underground service lateral.

Positively Shocking

Don't assume that you know where your utility company will allow you to install your service. The old location might no longer be up to code due to remodeling additions. Get the company's okay before you start planning your installation.

The service conductors are pulled through the mast and then pass through the meter (which records your electricity usage). Your utility connects the wires coming out of the weather head to the conductors (service lateral conductors or overhead wires) coming from the pole and the conductors to the meter, which the company usually supplies. Rules for installing an overhead service govern ...

➤ The location of the meter.

➤ Clearance requirements for the conductors.

➤ Securing and supporting the service mast and raceway to your house.

Your utility company determines where the meter will be located. The company's main concern is that the overhead conductors will be in a safe, unencumbered location. The clearance requirements for the conductors and the length of your service raceway (the mast and head) must take the following factors into account:

➤ The distance from the service mast to the utility pole

➤ The pitch of your roof

➤ Whether you're using an IMC, PVC, or steel raceway (conduit)

➤ The proximity of windows to the proposed location

Finally, the conduit or raceway must be secured properly so it doesn't loosen or bend. Any hole you drill for the raceway or its supports must be sealed so you don't get water leaks. Your local electrical inspector can give you more information about clearance requirements and securing your service mast.

Electrical Elaboration

You might not have a service head or service mast if a service-entrance cable is used. With this arrangement, the cable enters the outside wall of the house and is protected by a metal sill plate.

As you can see, this is a complicated process—and we haven't even gotten to the service panel yet! In case you were wondering, you can't attach any other cables (phone or TV, for example) to your service mast.

Going Underground

I'm a big fan of underground service conductors. New developments almost always have them, but a new house in an established neighborhood might not, depending on which side of the street it's located. If you're on the same side as the utility poles, consider burying your line. It will involve digging a trench at least 18 inches deep (check local requirements) and possibly doing so across a neighbor's property. A buried cable results in a much cleaner appearance, and there's no chance of it being damaged during severe weather, massive flooding notwithstanding. (The utility's power lines can still go down, however.) It's also out of the way when you have to set up ladders to paint or to work on your roof.

With an underground service, your utility company installs service lateral conductors (which may or may not have to be contained in conduit, depending on your local code). The conductors then enter the meter, via at least a short section of conduit, from underneath the ground. As with any outdoor wiring done by an electrician, you can save yourself some money by doing the digging and trenching yourself or by hiring it out to a less-expensive laborer than an electrician.

New Service/Old Service

Unless you're rewiring your entire house (which isn't likely unless you're doing a major remodel), you'll need to connect your new service panel with at least some of the existing circuits. This usually is not included in the cost of the service change. It really depends on the purpose of the upgrade. In addition, if you have an old fuse system, you will most likely replace the following (again, at an extra cost):

➤ The major appliance circuits

➤ The kitchen and bathroom circuits

➤ The water heater and possibly the furnace circuits

Existing branch circuits for lighting often can be left alone and simply tied into your new panel. (Each will have to be checked, of course.) If you are replacing an existing service panel on a three-wire grounded system with a higher-amp panel, you also will need to connect to the existing service. Only when you're completely rewiring your home can you ignore the existing service and its location.

Fuse Box Becomes Junction Box

When you tie a new service panel into some of the existing circuits in a fuse box, the circuits will originate in the service panel and be protected by the circuit breakers.

Individual cables, one for each of the retained circuits, are run from the service panel into the fuse box, where they are connected with the old wires. Because the new service panel is in a different location, the fuse box becomes, in effect, one large junction box. The door of the fuse box, which usually is free to open and close, is then screwed shut.

New Panel, Old Panel

If you have an existing service panel and a three-wire service with a grounding conductor, but you are installing a larger service, the new panel will take the place of the old one. New service conductors will be run from the utility's transformer, the existing circuits will be marked as the wires are removed from the old panel, and they will be reinstalled in the new, upgraded panel with new circuit breakers.

Anatomy of a Panel

A service panel must conduct electricity to individual breakers, must receive and route the current being returned through the neutral conductor, and must provide a grounding medium for the system. In a sense, it's the most powerful electrical device in your system. Remember, the conductors and cable that come into and leave your service panel include ...

➤ Two hot conductors.

➤ One neutral conductor.

➤ One grounding conductor that originates in the panel.

The two hot conductors energize the panel—and thus the breakers—via two *hot bus bars,* which are located in the center of the panel. The *black or red outbound wires* are connected to the circuit breakers that clip or slide into the hot bus bars. These wires supply the current to electrical loads throughout your house. The neutral wires are connected to the neutral bus with setscrew terminals. A grounding bus bar connects the various grounding conductors from the circuits to the panel's main grounding conductor. The grounding bus bar is bonded to the neutral bus bar. This is the only place the neutral and grounding conductors are tied together.

In addition to individual breakers, most service panels have a single *main service disconnect* in the form of an individual breaker or a series of high-amperage breakers connected together. The code requires that you be able to shut the entire panel down with a maximum of six hand movements. (That is, the panel can't need more than six switches or breakers to disconnect all of your home's electrical equipment.) An old panel might require up to six moves to shut everything down, but new service panels all have a single main shutoff, as previously described.

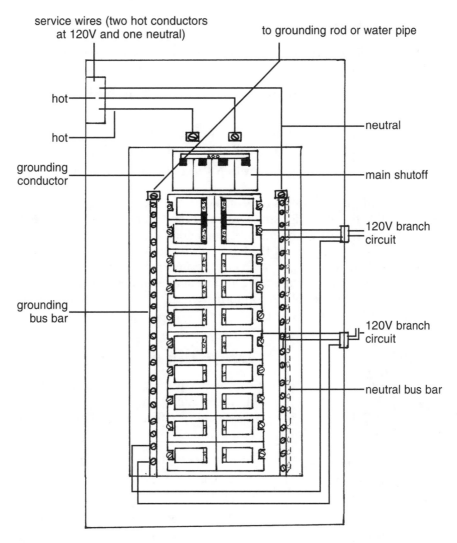

Inside of service panel.

The code requires that this main disconnect be as close as possible to the service conductors' point of entry into the building. In other words, you can't bring the service conductors into one corner of your basement and then install the service panel 15 feet away while exposing the conductors. (Certain exceptions do apply.)

Location Is Everything

The code is a little particular about where you can situate a service panel. In addition to being as close as possible to the service conductors' entry point, a service panel …

➤ Must have clear access to it (a minimum of a 30-inch wide by 36-inch deep uncluttered space).

➤ Cannot be installed inside cabinets or above shelving or any other encumbrance. It also cannot be installed in a bathroom.

➤ Must have a working space with 6½ feet of headroom around it.

In other words, the panel has to be in a clear and accessible area, and it must be readily visible to anyone looking for it. You can't store your skis or bikes in front of it, and you must be able to open the panel door a full 90 degrees until it's flat against the wall. Section 110-26(d) of the NEC calls for some illumination to be provided around the panel so that you or an electrician can see what you're doing if you have to access the panel. Although the code doesn't detail how much lighting you need to supply, you want to be able to comfortably read the panelboard (the breakers) and the *panelboard directory* inside the door.

Ask an Electrician

A **panelboard directory,** or **circuit directory,** is the formal name for your list of circuits. (I discussed this previously when creating a circuit map.) Every electrician makes a directory when installing a service panel.

Grounding Your Panel

A big safety advantage in newer service panels over old fuse systems is the fact that they're grounded. The code is very specific about grounding procedures including …

➤ The size of the grounding conductor.

➤ What is and is not an acceptable grounding electrode.

➤ Bonding requirements and the use of clamps and bonding bushings.

A service panel is grounded twice when possible: once to your water pipes (assuming they're metal and not plastic) and once to a grounding rod buried in the ground. When a grounding rod cannot be used due to soil conditions, a length of copper wire can be buried directly in the earth (a minimum of 2½ feet deep) or encased in concrete at least two inches thick that has direct contact with the ground (usually a foundation).

There also are other means of grounding the system (depending on your soil and house construction). The materials allowed, their installation, and their dimensions are all spelled out extensively in Article 250 of the code. Electricians know this stuff by heart because they use it every day. You, on the other hand, will never use it again, so consider calling an electrician for your panel installation.

Breaker, Breaker

Circuit breakers, along with fuses, are known as *overcurrent protective devices*. They protect you, your electrical equipment, and your wiring. They are matched to the ampacity of a circuit's conductors, and they shut the current down if there's an overload or a short-circuit. Breakers are clearly marked (15, 20, 30 amp, and so on) and must be used accordingly. You can't stuff a 20-amp breaker into your panel box and use it on a 15-amp circuit, especially if the circuit only has 14-gauge wire. This is a recipe for fire because you'll be allowing the wire to carry more current than it's designed to carry, and the breaker won't sense the problem and thus won't trip.

Positively Shocking

There's a good chance your local code won't allow you to install your own service panel. Some codes do not allow you to do almost any electrical work. On top of that, local inspectors will really scrutinize any homeowner-installed panel, as they should.

Breakers are either single pole or double pole (sometimes referred to as a two-pole breaker). Double-pole breakers are used for 240-volt circuits and draw power from each of the 120-volt hot wires entering the panel. A double-pole breaker can be either a single device or two single-pole breakers tied together so they'll both shut off at the same time.

Subpanels—a Real Convenience

Subpanels are smaller versions of your main service panel. They serve a couple of purposes:

➤ They provide proximity to circuit breakers, especially in large homes, so you don't have to access the main panel some distance away.

➤ They can expedite the wiring of a large house because the electrician only has to run one feeder cable from the main panel to the subpanel instead of running cable for every circuit separately the same distance.

A subpanel on the second floor of a large three-story house, for example, enables you to control the circuits on the second and third floors. Like your main panel, you have to follow a few rules regarding the location of a subpanel:

➤ It cannot be installed in a bathroom.

➤ It cannot be installed in a closet.

➤ The subpanel must be in an accessible, visible location.

243

Bright Idea

Don't like the looks of a sub-panel on a finished wall? As long as its location is clearly marked, you can cover the panel with a painting or a cabinet door. You cannot block access to it with furniture or other large encumbrances, however. You're better off leaving it visible, in my opinion, but you can paint it to match the wall, making it more visually tolerable.

Subpanel Alternatives

Before installing a subpanel, consider its necessity. Certainly, it's a convenience in the event of a tripped breaker, but is it worth the expense of purchasing a second panel (albeit a smaller one than your main panel)? The convenience factor aside, what you're really comparing is the difference in cost between running individual circuits all the way back to the main panel and running a feeder cable to the subpanel, the cost and installation of the subpanel, and the cost of running the branch circuits from the subpanel.

Some subpanels are installed because there aren't any breaker slots remaining in the service panel, even though the panel has the amperage to carry more circuits. In this case, your existing breakers can be replaced with a smaller version of a standard single-pole breaker. These go by different names, including …

➤ Slimline breaker.

➤ Peanut breaker.

➤ Mini breaker.

➤ Tandem breaker.

These breakers are half the thickness of a standard breaker, which enables you to fit two of them into a single breaker slot. Others have two breakers on one single-pole-size breaker.

Subpanel Considerations

The most logical locations for a subpanel in a remodeled house include …

➤ In a new addition.

➤ In a garage workshop.

➤ Near an attic converted to finished space.

A small subpanel can hold as few as two single-pole breakers, and a large one can hold up to 42. If the appearance of a larger panel isn't an issue for you, go ahead and install the larger panel so you'll have it available for future use.

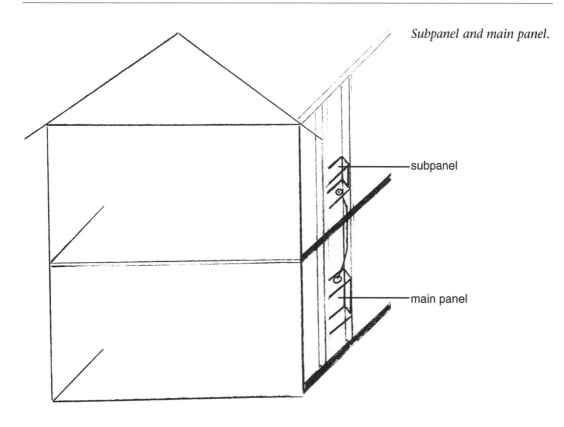

Subpanel and main panel.

subpanel

main panel

What's This Going to Cost?

According to *Today's Homeowner* magazine, the national average cost for upgrading an electrical service to a three-wire, grounded, 200-amp service is $2,264. For a little over $2,000, you get a safe, updated system and maybe a break on your insurance. If an electrician does the job, it will be done quickly and will be up to code. The electrician is responsible for meeting the code requirements. You'll have some clout because you won't (or shouldn't) be paying the final bill until the system has been inspected and passed by your local building department.

This chapter pointed out at least the major steps and considerations you'll face when upgrading your electrical service. This is well worth contracting out to an electrician, and I recommend that you do so, even if your local code allows you to install it yourself.

The Least You Need to Know

➤ If you still have fuses, it's a good idea to upgrade to new circuit breakers and a service panel.

➤ Even if you're a talented do-it-yourselfer, installing a new service is best left to an electrician, and local code might require that you do so.

➤ Whenever possible, consider burying your service connectors instead of running them overhead.

➤ Subpanels are a great convenience in many situations and are a doable project for amateur electricians.

Chapter 18

Adding New Circuits

In This Chapter

➤ Pen and paper first

➤ All the circuits you want

➤ Running lots of wire

➤ Marking everything

➤ Doing your neatest work

In some respects, doing a major rewiring job is easier than doing intermittent alterations. For one thing, you don't have to mess with tying into much of your old wiring because you'll be replacing it. Instead of updating the critical areas such as the kitchen and bathroom and just living with the inherent remaining limitations of the old system, you'll have upgrades everywhere. In other words, you'll be up to code (or mostly up to code) and be done with it.

Once your new service panel is installed (a job for an electrician), adding circuits is something most homeowners can do themselves. Even if you simply plan out and install the cable, you'll be saving a big part of an electrician's fees. Getting cable from the service panel to the device is the time-consuming part of the job. If nothing else, in a culture where we are increasingly disassociated from physical work, wiring your house can be a source of great pride and accomplishment.

Positively Shocking

Wireless connectors (often referred to as wire nuts, a trademarked name) come in different sizes to accommodate different numbers of conductors. Be sure you're using the correct size. Some will not form a tight connection if you twist the wire ends together first with pliers; others will. Refer to the manufacturer's instructions.

As with any alteration to your electrical system, you must have permits and pass an inspection. Some business consultants believe that a messy desk is the sign of an inspired, creative mind, but this isn't so with wiring. Freudian analysts might have a field day with electrical inspectors' obsession with neatness, but that's what they want to see, so don't disappoint them. Finally, plan your time. You don't have to do the entire house at once. You can do some of the work alone, but some is best done with two people involved. Check your calendars and pencil in—or punch into your personal digital assistant—a day that works for two of you.

Write Up a Plan

It's always a great temptation, at least if you're a guy, to dive into a project and improvise as you go. That's okay for standup comedy (at least when it works), but why add to your electrical labors when you don't need to? A pad of paper and a pen or pencil are still useful tools (even in the computer age) for visualizing your wiring. They can make the job easier and can save you some time by pointing out shortcuts and problems ahead of time. You might discover, for example, that a switch-controlled receptacle will provide light in a dormer bedroom more easily than trying to install a ceiling light.

You also might need a plan to get a permit, although not every building department requires one for electrical work. A plan will give you an accurate count of fixtures, devices, circuit breakers, and electrical boxes needed for the job as well as an approximate measure of needed cable. You don't want to go running back to a supplier because you're short two receptacles.

Power Everywhere

The whole point of a modern electrical system, aside from safety, is to have power, fixtures, and devices where you want them. You're only reading this chapter if you have an outdated system or if you're building your own home or addition. Newer houses rarely need circuits added unless they are being physically expanded or you're adding more power to an area such as an unfinished basement, a garage, or outdoors.

Don't underestimate your needs. If you've got the time, install all the receptacles a circuit can handle. You're already tearing up the walls, why find out later that the one wall on which you didn't install a device was where you could really use one?

Use the current code as a guide whenever practical for lighting and receptacle requirements. The only time you must follow it is during a major remodel when all the walls are open in an existing room and you're running new wiring or when you're adding on.

15 Amps or 20?

Most branch circuits for lighting will be 15 amps. Twenty-amp circuits normally are reserved for dedicated purposes. It's perfectly acceptable to use a 20-amp circuit for lighting, but use it judiciously because it can handle, for example, four more 100-watt fixtures than a 15-amp circuit. Great, you say, that means less wiring to do—at least until the lights go out. Then it might not be so great. You'll have that much less light to see by if an entire section of your house goes dark.

Twenty-amp lighting circuits work well when you have a large cluster of lights such as in a kitchen/hallway combination where you might have as many as 10 150-watt recessed fixtures. You also should consider a 20-amp circuit for your home office computer and peripherals. Check the rating of your copier, which could need its own circuit.

Plenty of Dedicated Circuits

You might have already updated your kitchen and bathroom wiring with 20-amp, GFCI-protected circuits (or opted for GFCI receptacles). Now you can expand and add other necessary dedicated circuits including ...

➤ An outdoor GFCI or two.

➤ A workroom circuit.

➤ A garage circuit.

Bright Idea

Even if you're not prepared to run all your new circuits at once, go ahead and drill any needed holes in basement or attic joist while you're drilling other holes. It might be too cold to install yard lights, for example, but you'll have access ready when you do the work in the summer.

Positively Shocking

If you have a fuse system, don't attempt to add circuits to it, even if you have ascertained that you have room to do so. This is a job for an electrician.

Roping the House

As a rule, you wire a house from the service panel outward. It's easier to drop a fish tape or chain downward through openings in the framing (and then pull up on the electrical cable) than to go the other way. After the cable has been run and the devices and fixtures are installed, you then do your final connections at the service panel. Working around a service panel doesn't have to be frightening, but you must be cautious.

Bright Idea

Mark the cable used for each circuit with a felt-tip marker of some kind so it can be identified. Also mark the inside of junction boxes. I'd even go one step further and mark the inside cover plates of every device. You can run a sheet of labels through your computer's printer for a fast, neat job.

Keeping the Inspections in Mind

An inspector will scrutinize your work more closely than an electrician's work. Make sure your cable is pulled tightly through wall and floor spaces and is stapled according to code requirements, which specify insulated staples or straps …

➤ Not more than 12 inches from a box or fitting.

➤ Not more than 4½ feet from each other when a cable is running along a stud or joist.

➤ Installed without damaging or denting the cable in any way.

You must have at least six inches of cable or conductors in each box from the point of entry into the box. Once attached to a device, the conductors should be neatly tucked inside with the hot and neutral conductors separated from each other. Overall neatness and professionalism go a long way toward satisfying an electrical inspector.

The Least You Need to Know

➤ Time spent planning your electrical additions can save even more time on the back end of the job.

➤ As long as you're making major changes to your system, add everything you can think of so you get it over with and only tear up the walls once.

➤ Mark all your new cable and devices as you go for future reference and so you don't lose track of the circuits.

➤ In addition to meeting code requirements, the job should be neat and professional-looking for its inspection.

Chapter 19

Kitchen Power

In This Chapter

➤ Special circuit considerations

➤ Wiring major appliances

➤ The more light the better

➤ The importance of ventilation

We take our kitchens very seriously. In years past, a kitchen was hidden away and was seen as a more utilitarian room used simply for the storage and preparation of food. These days, we dine, mingle, read the paper, and socialize in this room, which often has a family room directly connected to it. It's one of our home's biggest overall energy consumers, and it demands a lot of wiring, devices, and appliances.

Given the multitude of tasks and uses of a kitchen as well as the code requirements, you need to pay special attention to its circuits and the placement of light fixtures and receptacles. As previously discussed, you'll want plenty of task and ambient lighting. You'll also need a number of dedicated circuits for individual appliances. On top of that, some receptacles require GFCI protection and some do not.

A well-designed kitchen is a joy to be in and inevitably will become the hub of your house. Virginia Woolf said, "One cannot think well, love well, sleep well, if one has not dined well." I certainly won't claim that meeting the National Electrical Code in your kitchen will improve all these areas of your life, especially if you live mainly on microwaved hamburgers, but at least you'll get a better look at what you're eating.

The Well-Wired Kitchen

If you've ever lived with an outdated kitchen, you'll appreciate a modern one. Micro-wave ovens, food processors, and home-model espresso makers didn't exist in the 1920s and 1930s. One or two receptacles were plenty for the portable appliances available at the time. Trying to make do with your parents' or grandparents' wiring at the end of this century is an exercise in frustration, not nostalgia.

As you plan your kitchen upgrade, keep in mind any future remodeling. You might not be ready to replace cabinets, move walls, or upgrade appliances now, but you might in a few years. There's no point in going all out with your electrical changes if you have to redo them later. By all means, add the necessary receptacles, but think twice before adding those fancy light fixtures. All it takes is a different cabinet config-uration or the addition of an island to throw your lighting pattern askew. What if you move your electric range or refrigerator? If you run new circuits for them now in their present locations, you'll have to run them again later. Stick with the necessary work for your safety and convenience now. Of course, if you don't plan to do any ex-tensive future kitchen remodeling, go ahead with a full electrical makeover now.

Dedicated Circuits Everywhere

A kitchen is the home of the *dedicated circuit*. There are so many high-wattage devices here that too many on one circuit could cause it to trip; therefore, the code says these devices should be split up. Specifically, a kitchen should have individual dedicated circuits for the following loads:

Ask an Electrician

A **dedicated circuit** is used ex-clusively for a very defined load such as an electric dryer. The loads are considered by the code to be substantial enough to re-quire their own circuit. This avoids nuisance tripping and ex-cessive heat buildup, which could result if a substantial load shared the circuit with other loads.

➤ All built-in appliances including the dishwasher, range, microwave oven, trash compactor, and disposer

➤ Small, countertop appliances (food processors, toasters, and so on)

➤ Lighting fixtures (cannot be part of the preced-ing circuits)

A minimum of two 20-amp small-appliance circuits must be installed to bring a kitchen up to code. These same circuits can supply power to receptacles in ad-joining rooms including the breakfast nook and the dining room, but that's as far as they can go. You can't run your bedroom clock radio off a kitchen circuit.

Beyond the minimum, which is all the code addresses, you or your electrician must consider how your kitchen will be used. An espresso machine, for exam-ple, needs its own dedicated circuit if it runs at 1,200 or 1,500 watts to avoid tripping the breaker every time the espresso machine is running and you decide to use the toaster. You would still be legal with your two small-appliance circuits; you just wouldn't be practical given your intended use of the kitchen. If you have a lot of small appliances and use them regularly, consider adding a third or even a fourth dedicated circuit.

Small-Appliance Circuits and GFCIs

I already mentioned the danger of electricity around water and water pipes. An errant current will not hesitate to pass through you on its way to the ground through a water pipe. You also can get a shock or become electrocuted if you have contact with a receptacle and a range, a refrigerator, or a cooktop because they also are grounded. The code recognizes these dangers and steps in with GFCI requirements.

Specifically, the code requires that all small-appliance circuits used on countertops be GFCI protected. This includes any receptacles serving kitchen islands. The usual installation calls for the first receptacle on the circuit (the feedthrough receptacle) to be a GFCI type, which in turn protects the additional receptacles down the line. It also is acceptable to install a GFCI circuit breaker, although this is more expensive than a GFCI receptacle.

The key word here is "countertop." Other receptacles, such as one for the refrigerator or under the sink for a plug-in disposer, do not have to be GFCI protected. The under-the-sink receptacle cannot even be part of a small-appliance circuit; it must be its own dedicated circuit. Small-appliance receptacles must be installed so that any point on the back of the countertop is within 24 inches of a receptacle. Another way of saying this is that no two receptacles can be more than four feet apart. Every counter that's wider than 12 inches must have at least one receptacle.

Refrigerators merit their own discussion in the next section.

Electrical Elaboration

A **disposer** removes food waste by grinding it up so it will fit down your drain line. (This is the correct name for this appliance even though it's sometimes called a disposal.)

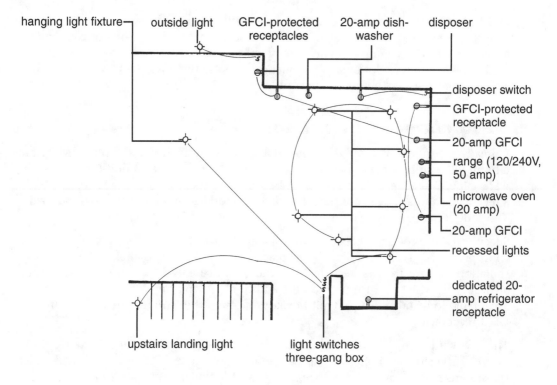

Typical kitchen wiring diagram.

Wiring the Fridge

Some modern refrigerators draw a surprisingly small amount of current for all the work they do. Although the NEC figures 1,200 watts in its load calculations for a refrigerator, some only require half that amount. It's apparent that it takes more electricity to produce heat than to maintain cold. The code allows a refrigerator to be supplied by its own 15- or 20-amp circuit, but it does not require a dedicated circuit. You can run a refrigerator off one of your small-appliance circuits, but many electricians recommend a dedicated circuit for the fridge.

Why? Remember, your small-appliance circuits are GFCI-protected. All it takes is one instance of nuisance tripping, and your refrigerator could be off for hours (or weeks, if you're out of town). In addition to losing a lot of food, you'll have to clean out some pretty rank stuff. A Sub Zero brand refrigerator (or other high-end refrigerator) requires a dedicated 15-amp circuit.

Two More Exceptions

The code allows two specific receptacles to be hooked up to a small-appliance circuit:

➤ A clock receptacle that only supplies power to a clock

➤ If a gas range is used, a receptacle to power the clock and the electronic ignition, if present

You can't run anything else on these circuits. A ventilation fan can run off the lighting circuit, unless it's big enough to rival your favorite fast-food place. In that case, depending on its wattage, you might want it on a dedicated circuit, too. As a rule of thumb, consider a dedicated circuit for any load over 1,000 watts.

Ranges and Ovens

An electric range requires a lot of wattage (12,000 isn't unusual). It also requires a 50-amp, 120/240-volt circuit, a specialized receptacle, and No.8 cable, which isn't cheap. You don't want to run this to one location, only to change your mind and decide you really want the range two feet farther away.

Disposers

Disposers are another kitchen toy few of us live without in newer homes. They're now installed in houses of all price levels. Oddly, they were not legal in New York City until August 1997, when concerns that zillions of pounds of ground-up orange peels, unfinished blintzes, and coffee grounds would clog the city's plumbing gave way to consumer demand, a desire to decrease waste in landfills, and health hazards associated with uncollected trash bags full of food waste. Local laws in New York say that disposers can be installed only by licensed plumbers and electricians, who need permits to do the work. New Yorkers can contact the Department of Buildings at 60 Hudson Street to get the application for permits.

Positively Shocking

Clock receptacles, which are recessed into the wall so a clock will fit flush against it, aren't commonly used for clocks anymore. Increasingly, they're used for microwave ovens in the kitchen. If yours is used for a clock and is part of a small-appliance circuit, however, you have to continue using it for a clock. The code won't allow you to plug a microwave oven into it.

Bright Idea

Buy the most powerful disposer your budget will allow. The installation costs are approximately the same for all models by a specific manufacturer. It's worth paying the extra $30 or $40 to avoid buying an underperforming model. Trust me on this one.

Disposers either are hard wired or are plugged into a receptacle under the sink with an appliance cord and plug. Some local codes require that a disposer be plug connected. A plug-in model is easier to work on because you can disconnect it from its power source quickly for servicing or replacement. This receptacle cannot be used for any other load, nor can it be part of a small-appliance circuit.

A disposer can have one of the following three types of "On/Off" switches:

➤ A standard single-pole switch located above the counter or under the sink.

➤ An integral switch built into the drain lid that, when twisted, activates the unit.

➤ A flow switch installed in the cold-water line and activated when the drain lid is twisted. With this type of switch, the disposer will not run until cold water is flowing into it. (This helps prevent grease buildup.)

Disposers are even available for septic systems. In-Sinkerator makes a complete line of disposers. You can see the company's products at www.insinkerator.com.

Electrical Elaboration

Disposers come with thermal overload protection for their motors. It kicks in and shuts the unit off if it's being operated in excess of its normal working capacity or load rating. This can happen when an item such as a bone becomes lodged in the disposer's blades. You can manually press a small reset button on the side of the disposer to allow the unit to operate again. You probably will have to move the blades manually with the manufacturer-supplied Allen wrench first if they're jammed.

Dishwashers

Some energy-conservation advocates don't like dishwashers and see them as hugely wasteful in terms of the electricity they demand, the amount of wastewater they produce, and the pollution caused by their detergents. There's even disagreement as to which part of a dishwasher's cycle demands the most energy: heating the water or drying the dishes after they're washed and rinsed.

Whether you agree with these arguments or not, just try selling a house without a dishwasher in it. They're wonderful appliances, and they come with a multitude of settings and options through which you can control some of their energy usage. To alleviate your conscience, you can decrease your energy usage by ...

➤ Letting your dishes air dry.

➤ Purchasing a dishwasher with energy-saving features.

➤ Only washing full loads (but not overloading).

➤ Buying a standard-capacity unit instead of a compact-capacity unit. (You'll do fewer loads, thus using less water and electricity.)

➤ Installing your dishwasher away from the refrigerator. The dishwasher's heat can cause the refrigerator to demand more power to keep its contents cold.

A dishwasher should be on its own dedicated 15-amp or greater circuit depending on the model. Portable dishwashers can be plugged into an existing countertop small-appliance receptacle.

Big Appliances: Stovetop and Range

Next to an electric furnace, an electric range is your home's largest user of electricity. It runs on a 120/240-volt circuit. (The 120 takes care of any lights and the clock.) Most units are freestanding with their own oven and stovetop and a cord-and-plug connection, but some homes have the two separated with a wall-mounted oven and a counter-mounted stovetop, both hard wired to a junction box.

A range requires a 50-amp, *double-pole circuit breaker* and No.8 cable. Cord-and-plug-connected ranges require a special range receptacle rated at 50 amps. If you're buying this receptacle, don't confuse it with a dryer receptacle, which is identical in appearance but is only rated for 30 amps! Dryer and range receptacles feature four-wire construction (two hot leads, one neutral, and one grounding conductor). Note that, prior to 1996, a three-wire receptacle was used for these installations, and the grounded conductor (the white or neutral) was used to ground the appliance. If you have an existing, pre-1996 dryer or range circuit without a green or bare copper grounding conductor, you might install a new appliance using the existing method (ground the appliance frames to the neutral). If you can conveniently run a new circuit, I'd recommend that you do so.

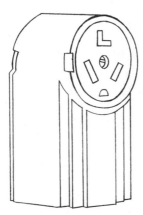

Stove receptacle.

Keep or Replace Your Appliances?

Kitchen appliances can be expensive. A super-silent dishwasher with multiple settings can run as much as $1,000. The same is true of refrigerators. You have to establish your own criteria for replacing your existing appliances. The following are common reasons for replacing appliances:

➤ Current appliances are out of style, are the wrong size, or don't match remodeling plans.

➤ There are age and repair issues.

➤ You desire energy savings.

If you're simply updating your kitchen's wiring and your old appliances are still in working order, consider keeping them until you do a full-scale remodeling. New appliances might not fit in your existing cabinet and counter configurations. (Depth and side clearances might be different depending on the age of your current appliances.)

Lighting the Way

Your kitchen can feature a wide variety of lighting: small fluorescent or low-voltage strip lighting over the counters, recessed ambient lighting in the ceiling, and hanging fixtures over the eating nook. You'll never exhaust the possible fixture choices and combinations available for a kitchen. You should provide plenty of work light for food preparation and ambient lighting for everything else. A kitchen is a main gathering point for many homes, so you want it to be comfortable and appealing.

Branch circuit kitchen lighting typically is 15 amps with three-way switches at each entrance (and a four-way switch when called for). The lights must be on a separate circuit from the kitchen appliances.

Work Lights

Undercabinet lights usually are hard wired with either a wall switch or individual switches on the fixtures themselves. You want to install plenty of them over any counter wide enough to use for food preparation. Installation options in or near upper cabinets vary and must be coordinated with your cabinetmaker/installer if you're wiring a new kitchen or doing a kitchen remodel. Installation options include …

➤ The rear wall under the back side of the cabinets.

➤ Attached to the bottom of the cabinets themselves.

➤ Inserted in a built-in recess at the back of the cabinets that keeps the fixture itself out of sight.

Positively Shocking

Old-house wiring might not be suitable for some fixtures. Some new fixtures, for example, generate more heat than the insulation on old wiring can tolerate. Some fixtures are even marked for high-temperature wire only. If you're uncertain about the compatibility of your proposed new fixture and your wiring, consult an electrician.

In addition to fluorescent fixtures, small track-lighting systems also are available for undercabinet installations. Note that these fixtures often use halogen bulbs that create more heat buildup than their fluorescent counterparts. Overhead fluorescent fixtures installed in the ceiling will provide plenty of ambient lighting and will require less wiring than individual recessed or hanging fixtures. Another approach, if you're trying to keep the amount of ceiling hole cutting and patching to a minimum, is a full-size track light that only requires one electrical box to bring the wiring to the fixture yet can hold multiple lamps.

Accent Lighting

Kitchens aren't often associated with the ambiance of a living room full of artwork. Nevertheless, some accent lighting can add to their appeal. Small wall washers can be installed at the top of cabinets to highlight the area between the cabinets and the ceiling. A couple of low-watt sconces can double as great night-lights.

Lots-o-Switches

You don't have to have all your kitchen lights go on at once in one huge sunburst. You can divide up the loads with as many switches as you want. This not only controls the light level but also your energy use. Dimmer switches can be used for both incandescent and certain fluorescent fixtures. Keep in mind that the more switches

you have per individual or group of fixtures, the more wiring you'll have to do. In open walls with the studs exposed, this isn't much of an issue, but it can be tedious when working around finished plaster or drywall.

Ventilation

We associate kitchens with the wonderful smells of warm cookies, homemade bread, and roasting turkeys. Unfortunately, we also get burned eggs and the various contents of pots and pans that overflow onto hot burners. Somehow these smells seem to linger a lot longer than the smell of hot-out-of-the-oven buttermilk bread.

A kitchen exhaust fan removes odors, steam, and heat buildup from cooking. If the steam isn't carried outside, it ends up on your kitchen's ceiling, walls, and cabinets, which means more cleaning and repainting. A variety of models and sizes are available, from a small ductless fan to restaurant models with huge range hoods (that accompany huge, restaurant-size stoves).

You Have a Choice

The two main categories of exhaust fans are ducted and ductless. A ducted fan is attached to a system of metal ductwork that vents to the outside of your house. A ductless fan contains an activated charcoal filter that captures the particulate from cooking and recirculates the air back into your kitchen. Ductless models are better than nothing, I suppose, and typically are installed in remodels where running ductwork would be prohibitively expensive. Given a choice, you're better off with a ducted fan.

Up and Out, Down and Through

Kitchen exhaust fans situated over a range top will pull air in and duct it outside either through a wall or through the roof. A downdraft cooktop, often used in kitchen islands, has a built-in blower that exhausts through the base of a cabinet or through its back. Installing a fan means more than just electrical work. The job also will include …

➤ Cutting away an existing cabinet or designing your new cabinets to accommodate the fan.

➤ Running the ductwork through the wall.

➤ Cutting an opening in an outer wall or through the roof to install the duct cap.

➤ Caulking and sealing the duct cap.

As with bathroom exhaust fans, kitchen fans come in a range of prices and quality.

Bigger Is Better

I'm a big believer in getting steam and odors out of a kitchen quickly rather than using a barely adequate fan and listening to it drone on and on while the kitchen still smells like sautéed onions. Nutone, a major manufacturer of exhaust fans, offers quite a line of models:

➤ The HCH5200 is a high-performance range hood with twin blowers (independent controls) and a stated capability to remove and trap airborne oil.

➤ The SH-1000R, the Continental Slide-Away Range Hood, slides away when not in use.

➤ Several models are mounted outside on the roof or on an exterior wall, so you never hear motor noise.

It's easy to purchase the first fan you come across at your electrical supplier or home-improvement store, but a little research will pay off in the long run. Remember, it's better to vent your kitchen quickly with a powerful fan and shut it off sooner than to vent it slowly with a smaller fan. Some larger fans can be noisy, and you might consider mounting these models remotely on the roof or exterior wall.

Electrical Elaboration

Some of the neatest exhaust fans I've seen were installed in a 1920s apartment house in Seattle. Each was a small-diameter, circular-shaped unit mounted through a hole cut in the window glass. A flush receptacle had been wired in the window jamb. The fans turned on automatically when the metal door on the inside, which kept cold air out, was opened.

Wiring Concerns

Kitchen wiring is fairly straightforward. Every dedicated circuit is directly wired to a circuit breaker (although it might be at a subpanel or pass through a junction box between the load and the breaker). All 20-amp small-appliance circuits must use 12/2 cable. Lighting and other 15-amp circuits can use 14/2 cable, although 12 is acceptable as well. I recommend that all circuits except lighting circuits be wired in 12/2. It helps keep things simple and makes it easy to install any new appliance that comes along.

The Least You Need to Know

➤ Kitchens are full of dedicated circuits for just about every fixed appliance.

➤ Small-appliance circuits cannot be connected to any built-in appliance.

➤ The kitchen is one room that can use plenty of lighting fixtures; the huge variety of styles gives you plenty of interesting choices.

➤ Your kitchen deserves a good-quality, high-capacity exhaust fan.

Now THERE'S a bad idea...

Chapter 20

Bathroom Wiring

In This Chapter

➤ Installing a GFCI

➤ Getting the steam out

➤ Extra heat for extra comfort

➤ Lighting choices

➤ Tile walls

Like kitchens, bathrooms have evolved into more comfortable (and more costly) rooms. Although you could find fancy, fully tiled bathrooms in high-end homes at the turn of the century, contemporary bathrooms feature decorator fixtures and lighting even in modest houses. From an electrical standpoint, modern devices and wiring cover everything from heated floors to heated, steam-free mirrors. Even the Romans, who were famous for their public baths, didn't have it this good.

Bathroom remodels and additions are among the most expensive per square foot (usually second only to kitchens). You can choose from myriad lighting options, more than one way to bring in needed heat, and several lines of exhaust fans. A perfectly serviceable and attractive bathroom can be had without sacrificing the kids' college funds, unless you decide they should work their way through. Then you can start thinking about a fireplace, a Jacuzzi, and a built-in steam room.

GFCI Is a Must

The key code-required device in a bathroom is a GFCI receptacle. One could argue that it's more critical here than in the kitchen, given the constant proximity to water or water pipes while using a bathroom. The code is very specific about bathroom GFCIs:

➤ The GFCI must be located on a wall adjacent to the basin; it cannot be installed face-up on the countertop surface surrounding the basin.

➤ The GFCI must be on a 20-amp dedicated circuit. Section 210-11 (c)(3) allows more than one bathroom to be connected to this single 20-amp circuit.

➤ The dedicated 20-amp circuit cannot supply any load other than the GFCI unless the circuit is only supplying a single bathroom, in which case it can supply other loads.

Given the ever-present use of hair dryers, shavers, and other grooming toys, you're usually better off running separate 20-amp circuits to each full bathroom. This also enables you to run the lights of the GFCI. As a 20-amp circuit, it will require 12/2 cable.

Oddly, the NEC defines a bathroom (Article 100) as "an area including a basin with one or more of the following: a toilet, a tub, or a shower." A basin alone in a room is not a bathroom, despite the potential hazard it might pose for the user of a close-by receptacle. In some turn-of-the-century homes, basins were installed in some bedrooms and are still present today. The code does say that any receptacle within six feet of the countertop of a wet bar must be GFCI protected, and your local inspector might abide by this standard when it comes to stand-alone basins.

Bathroom wiring plan.

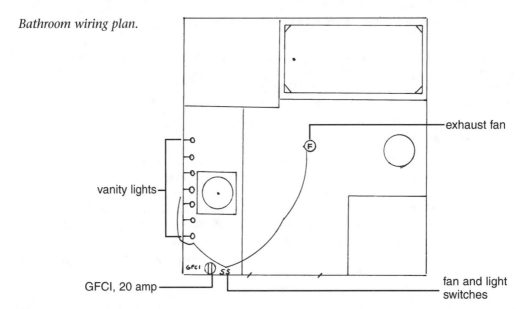

Your bathroom's wiring doesn't end with the GFCI. Other fixtures include ...

➤ Exhaust and ventilation fans.

➤ Lights.

➤ Electrically heated floors.

➤ Heated towel bars.

➤ Baseboard- or blower-style heaters.

➤ Heated, fog-free mirrors.

➤ Ceiling-mounted heat lamps.

➤ On-demand water heaters for the tub.

At this rate, you could get your entire electrical education just by working on your bathroom!

Installing the Fan

Bathroom exhaust fans are well worth the effort to install if you don't already have them, which often is the case in older homes. They are required in new bathrooms. All that steam from the shower has to go somewhere, and if it isn't vented outside, it lands on your walls. This usually means mold, mildew, and more frequent repainting than you really want to do. Let's face it: No one wants to repaint a bathroom more often than necessary. A good ventilation fan will help extend the life of your paint and will keep your bathroom fresher.

You have several choices of exhaust fans:

➤ Vent-only

➤ A vent and light combination

➤ A heat, vent, and light combination

A vent-only fan gives you the most options regarding size and power. Vent and light combinations run from basically useless, barely adequate units to more substantial models. A heat, vent,

Positively Shocking

Code requires that the front edge of your electrical box be flush with the finished wall when combustible material (framing) is involved. This way, the cover plate will completely conceal and protect the device and wiring. You can't have half your box flush against the drywall and half not flush against tile. You are allowed ¼-inch of play with noncombustible materials, however.

Positively Shocking

Never install a switch or a receptacle in the tub or shower area. The only time a switch is acceptable is when it's part of a manufactured tub unit—for example, one with a built-in stereo—and it's been approved by Underwriters Laboratory or another testing facility.

and light combination will not heat the room as efficiently as a separate wall heater will. After all, heat rises. When it starts out at the ceiling, it's at a disadvantage.

In addition to ventilation options, you also can choose between ceiling-mounted and through-the-wall fans.

How to Size It

Exhaust fans differ in a number of measurements (physical size of the unit, amp rating, lamp watts), but the most critical one is cubic feet per minute (CFM). This tells you how much air the fan will pull out of your bathroom and the rate at which it will do so. CFM ratings determine how many square feet of bathroom a particular model is suitable for. My experience has told me to ignore the ratings and to install larger-capacity fans because I think it's more important to clear the steam out faster and to shut the fan down sooner.

If you've ever lived in a new house, I think you'll see my point here. Unless you specify otherwise, your new bathrooms will come with the smallest fan allowed by square-footage ratings. Its main function is to make noise so you'll ignore all the steam that's still floating around and fogging the mirror. There's the argument that a larger fan will pull out the heat in the room too quickly to be replaced at a comfortable level, but this hasn't been my experience because a large fan is used for a shorter amount of time than a smaller fan. Besides, if you have a variable-speed control, you can always slow down the fan's speed.

Given that there's little difference in labor cost between installing a fan rated at 100 CFM and 200 CFM (or even 300 CFM), I'd go with the larger unit. The cost of your ducting material and outside cap, however, most likely will be higher with a larger model.

Roof or Wall Vent?

The best all-purpose bathroom exhaust fan is installed on the ceiling, where steam naturally rises. As a compromise, some fans are designed for through-the-wall installation, which requires no ducting for most walls. (Nutone's model 8870 can accommodate walls 5 to 10 inches thick.) These fans are installed near the tub or shower for best use, and they must be on a GFCI circuit when inside the tub or shower area.

Ceiling models come in round, square, and rectangular shapes. There also are ductless fans made with activated charcoal filters for odor removal and keeping air fresh, but they do not remove any steam. I wouldn't recommend them for bathroom use, although they might be acceptable for half-baths or powder rooms.

So Many Different Switches

It probably wouldn't be America if we didn't have at least a dozen or so different fan switches from which to choose. In addition to a standard toggle, you also have the following options (and others not listed):

➤ Variable-speed controls (that enable you to choose any fan speed)

➤ Timer switches (15-minute, 30-minute, 60-minute, and so on)

➤ Multigang boxes with separate controls for the vent, light, and heat

➤ Vent and bathroom light controls on one switch so they both go on when the light is switched "On"

A timer switch gives you a lot of options. You can manually turn the fan on and off, or you can keep it running for a preset amount of time after you've left the bathroom (a good idea after particularly long showers).

A vent-only fan or a vent and light combination will run off a 15-amp circuit with 14/2 cable. A heat, vent, and light combination fixture most likely will require a 20-amp dedicated circuit, depending on its amp rating. This is considered an "electric space-heating unit," and it falls under a different section of the code than a fan alone.

It's Cold in Here

I don't think I've ever been in a house where at least one bathroom, usually the biggest one, couldn't stand to have some supplemental heat during the winter. An engineer might argue that it's unnecessary, that the room is adequately warm according to the calculations and so on, but human perception is far different from theoretical constructs. If you feel cold, you are cold, and that's all there is to it. You can adjust to it by adapting to a different degree of comfort, or you can put in extra heat. Because few of us want to relive the days of the Vikings, we usually choose more heat.

Electric heat is available for bathrooms in several flavors:

➤ Baseboard heaters

➤ Forced air, in-wall heaters

➤ Heated floors

➤ Heat lamps

➤ Heat, vent, and light combination fixtures

The voltage requirements vary depending on the model of heater you choose. Some will be 240 volt; others will be 120 volt. Because you're only supplying supplemental heat, a smaller, 120-volt model probably will suffice in most situations, but a 240-volt model gives you more wattage capability for the same amount of wire. The only difference in cost is the double-pole breaker rather than the single-pole breaker.

All of these heaters represent *zone heating*. That is, they are independent heating units whose purpose is to heat a limited area or a single room. A thermostat can be installed on the unit itself or in a remote location on a wall.

Baseboard Heating and Forced Air

Baseboard heaters take longer to heat an area because of their design. An element warms up as an electric current passes through it. The element's metal plates or fins direct the heat outward into the room. Cold air under the heater is drawn upward and is heated by the element. Once the element gets sufficiently warm, as determined by the thermostat, baseboard heating is very comfortable. If you take a shower in the morning, you'll want to turn on a baseboard heater as soon as you wake up so it can start heating the bathroom.

A forced-air unit will give you more immediate heat, which might be more appropriate for most bathroom usage. A small built-in fan blows room air over a heating element. There will be some fan noise, however, unlike the complete silence of a baseboard heater.

Positively Shocking

Fire departments hate space heaters and zone heaters because they're subject to misuse. Every winter, a huge percentage of house fires in cold climates are caused by independent heating units. Do not drape clothes or towels over a zone heater. Keep combustible materials (hair spray, for example) away from them. To clean, vacuum the fins or grills periodically to remove dust.

Heated Floors

This technology goes back to the Romans, who built fires under the stone and marble floors of public baths. We have advanced the technology some, but the principle remains the same: Heat rising from the uniform surface of a floor will provide the most even heating available. As an added bonus, a heated bathroom floor is easy on bare feet.

Also known as underfloor warming systems, heated floors typically require the installation of some kind of heating pad by embedding it in the tile's mortar. Amp ratings vary, but it's safe to assume you'll need a dedicated GFCI circuit. The NEC rules in Section 424-3(b) state that any circuit used for fixed electric space heating must provide circuit-breaker protection that is not less than 125 percent of the heating unit's ampere rating. If a unit's rating is 13 amps, for example, the circuit's breaker must offer 16.25 amps of protection. In this case, it will require a 20-amp breaker to meet code.

Installing a heated floor is a very involved job. It usually is done during new-house construction or when adding a new bathroom.

Heat Lamps

A heat lamp is an overhead fixture with a large, high-wattage bulb that provides only nominal comfort in the form of radiant heat and only if you're standing directly underneath it. These lamps run off a standard 15-amp branch circuit. A heat lamp is okay if you have no other convenient supplementary heating options for your bathroom, but I wouldn't expect to linger after stepping out of the shower or bath.

Some Codes Don't Like Electric Heat

Check with your local building department regarding its codes pertaining to electric heat. Some might require a certain level of insulation and possibly the installation of an insulated window once you start talking electric heat. They like to keep the demand for electricity down, and they prefer that you stick with your furnace. This policy might apply only to whole-house electric heating, or it might apply to individual rooms as well.

More Bathroom Toys

They might not be common, but electrical conveniences in bathrooms don't stop at heating and ventilation. You also can install …

➤ Heated towel racks.

➤ Heated mirrors.

➤ On-demand water heaters for tubs.

These all go above and beyond normal consumer needs, but they certainly awaken consumer wants. If you've got the wire, the time, and the budget, you can find plenty of electrical projects for your bathrooms.

The Luxury of Warm Towels

Once a feature almost exclusively found in certain high-end resorts and hotels, electric towel warmers can be a real treat. Designs vary from a basic towel bar warmer to more elaborate nickel-plated, freestanding units. Both hard-wired and plug-in models are available.

Although warmers are available that operate on hot circulating water, the most efficient have a low-wattage heating element. Models with timers are the most convenient because they will warm up the towel before you need it and will turn off the heat at a preset time. Sophisticated models offer several on/off cycles so that everyone can have a warm towel regardless of the time he bathes. Towel warmers are considered a single fixture and run off a 15- or 20-amp circuit.

Fog Be Gone

If you've gotten as far as heated towels, you might as well consider a heated mirror, too. Several manufacturers offer defoggers that are simple to install and draw very little current. A defogger can run off an existing 15- or 20-amp branch current for the lights or fan if the circuit can carry the added load.

The defogger is a thin heating pad that attaches to the wall behind your bathroom mirror. If connected to the light switch, it automatically turns on when the lights go on. An electrical box is needed behind the mirror to wire the defogger, and it does not have to be on a GFCI circuit. Some mirrors, on the other hand, are designed for use in a shower and must be GFCI protected.

On-Demand Hot Water

These units are more popular in Europe because of their efficiency and relative ease of installation. An on-demand unit heats water as it passes through the unit. In theory, you never run out of hot water because you're not depending on a reservoir of water in a tank. Available in both gas and electric models, each unit requires a source of power (gas or electricity) and must be directly hooked up to a fixture's plumbing with shutoff valves available.

I don't know when, if ever, these units will catch on in the United States for residential use. They're a useful idea for very specific purposes, however, such as at campsites where it's impractical or impossible to run a gas line and maintain a hot-water tank.

Bathroom Lighting

Bathrooms are comparatively small rooms, but you have quite a selection of lighting options for them. You want enough lighting for both grooming and reading, whether you're reading the instructions on a bottle of medicine or the *New York Times*. You also might want accent lighting around the tub, a light over the shower, or maybe a built-in night-light.

Bathrooms have become decorator showcases with prices to match. Fixtures and even cover plates can be as costly as you want them to be. In the end, remember that you don't live in this room (certain teenagers and would-be models being the exceptions), so consider how badly you really want that high-priced fixture. From a wiring standpoint, you are looking at a 15-amp circuit with a minimum of 14/2 wire.

Incandescent Fixtures

The most common location for primary lighting fixtures in a bathroom is over the basin or vanity. This gives us plenty of lighting for grooming and the usual facial scrutiny we all give ourselves from time to time. The size of your mirror will determine the best choice of fixtures. A small mirror on a single medicine cabinet, for example, can easily be illuminated with wall fixtures on each side or with a single- or double-lamp fixture overhead.

Larger mirrors really call for a bathbar, which is a multilamp fixture that comes in a wide variety of sizes, finishes, and lamp covers or globes. Sixty-watt lamps provide plenty of light. One advantage of a bathbar is that it only requires one electrical box yet gives you multiple lamps, so you're only wiring one fixture. Another choice is to install theatrical lighting strips all around the mirror, giving you the same lighting that actors get for applying their makeup.

Recessed lighting is the most common choice for bathroom ceiling fixtures. These fixtures provide plenty of ambient light yet are unobtrusive, always an important consideration in small rooms. Both recessed fixtures and bathbars are available with halogen lamps as well.

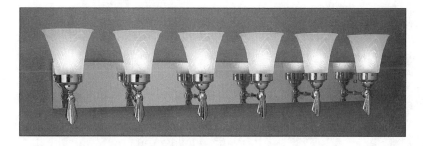

Six-light bath bracket fixture.

(Progress Lighting)

Broadway lighting for the bathroom.

(Progress Lighting)

Four-light bath bracket fixture.

(Progress Lighting)

271

Bright Idea

Do you have more than one bathroom needing new or up-dated fixtures? Do your budget a favor and prioritize a little. Put the best fixtures in the master bath and powder room, the next-best fixtures in the guest bath, and the most modest in the children's bathroom. (They won't be impressed with anything fancy anyway.)

Bright Idea

If you have a large house and a number of kids, you might consider installing a washer/dryer combination in an upstairs bathroom in addition to your regular laundry room. These models are designed to fit in spaces approximately 24 inches wide, about 28 inches deep, and 6 feet in height. The usual 120/240–volt circuit is required.

Tub and shower areas require a *damp-location fixture*. These come equipped with gaskets to keep moisture from seeping into the fixture and the electrical box.

A Possible Laundry Room?

Some homeowners are designing their bathrooms to double as laundry rooms. This is an idea worth considering if you have the room. A ground-floor bathroom is an obvious location, but a second-floor bathroom can be used as well with a few modifications including ...

➤ Reinforcing the floor joist.

➤ Installing a concrete fiberboard as an underlayment.

➤ Installing a floor drain.

Washing machines bounce around occasionally due to unbalanced loads. You don't want the floor to shake so much that the ceiling below it cracks. A floor drain will take care of any leakage from broken hoses. Floor drains work best in a concrete floor of some kind rather than in a plywood underlayment. As far as your wiring is concerned, your washer and dryer will be connected the same as they would be in a separate laundry room:

➤ A 30-amp, 120/240-volt circuit for the dryer using 10/3 cable

➤ A 20-amp, 120-volt circuit for the washing machine receptacle and a second receptacle for other laundry equipment.

This installation involves plumbing as well, so you must coordinate various trades, depending on how much of the work you do yourself. Although the dryer will be vented, it's important that the bathroom have a strong ventilation system to compensate for the added moisture from doing laundry.

Electrical Elaboration

In 1993, the U.S. Patent Office issued a patent for Toilet Landing Lights, which apparently are for people who don't know about night-lights for their bathrooms. Waterproof, indirect lighting is installed under the rim of the toilet. A switch attached to the lid activates the lights when the lid is down. If the lights are off, let's just say you won't be walking into the situation entirely unprepared.

The Least You Need to Know

➤ Remember, the NEC requires a GFCI in every full or partial bathroom.

➤ The purpose of a bathroom fan is to get the steam out fast, so think bigger rather than smaller when you shop for a fan.

➤ Even if your bathroom has a heat register, a supplemental heater, such as a baseboard heater, might be welcome on those cold winter mornings.

➤ If you're so inclined, look into installing towel warmers and heated, fog-free mirrors in at least one of your bathrooms.

➤ You don't need to overwhelm your bathroom with light, but make sure you supply enough for grooming and reading.

273

Chapter 21

The Great Outdoors

In This Chapter

➤ Appropriate light levels for outside

➤ Planning the locations

➤ Conduit or underground cable?

➤ Choosing the fixtures

➤ Security and comfort

In the past, outdoor lighting consisted of a porch light over each entrance to the house and maybe a light over the garage. Really fancy lighting meant putting out the Polynesian-chic kon-tiki torches around the patio and lighting the wicks. Landscape and security lighting, however, now provide a new nighttime perspective throughout the year.

As with indoor fixtures, you have many types and styles to choose from when you start wiring your yard and house exterior. This leads to the issue of when enough is actually too much. The notion of light pollution is not a new one, but it is getting more attention from local governments as they pass ordinances about outdoor lighting. Even if your local government doesn't have such ordinances, you want to stay on good terms with your neighbors. That's not going to happen if you install megawatt mercury vapor lights that make your yard as bright as a used-car lot at night.

Positively Shocking

Remember that your yard already has things buried in it. Water pipes, gas pipes, and phone, cable, and electrical lines are all possible obstacles. If you're not sure of their presence or location, call the respective utility companies. They can assist you in locating pipes and cable on or near your property.

Outdoor lighting has its own code requirements. Weather and moisture are big influences on wiring practices. Lamps and wire connections must be kept dry, any cable has to be rated for underground use, and individual conductors must be run inside conduits. With a little planning, and by following the code, you'll have an inviting and safe outdoors.

Light Up the Night

The main reasons for outdoor lighting are convenience, safety, and security. Without lights over or near our entry doors, we all would be fumbling for keys or tripping over the steps. If someone is thinking of burglarizing a house, a poorly lit home makes an easier target than one with adequate lighting. Unlike certain prehistoric fish living thousands of feet below the ocean's surface, we don't do well in the dark. "Let there be light" has become a homeowner demand!

A well-planned outdoor lighting scheme meets several criteria:

➤ It provides sufficient light for the job at hand (lighting a sidewalk or porch, for example) but doesn't overwhelm with light.

➤ Fully shielded fixtures direct the light to the area needing illumination instead of producing glare from misdirected and wasted light.

➤ The fixtures are carefully positioned to affect a targeted area with uniform lighting without affecting the neighbors' property.

➤ It avoids an unequal mix of very bright areas with adequately lit areas, because the latter can appear too dark and be less safe to maneuver around.

It's easy to get carried away with outdoor lighting (that is, if you like to dig ditches for the cable), but too much of a good thing not only can cancel out its benefits, it also is illegal in some municipalities.

Light Pollution

This isn't the pollution that caused the Cuyohaga River in Ohio to catch fire some years ago (most people from Ohio know this story), but it's an issue nevertheless. In addition to the aesthetic issues involved with excessive outdoor light, there also are legal and safety concerns. Local governments are increasingly discovering *light pollution* (namely, *sky glow* and *uplight*) and are beginning to enact rules concerning it.

The town of Kennebunkport, Maine, came up with a lighting ordinance six articles in length. (Maybe local resident and former president George Bush was annoyed by one of his neighbors' yard lights.) The town's reasons are stated here:

Ask an Electrician

Light pollution can be defined as an excessive amount of outdoor lighting that is improperly planned and installed, which produces inordinate glare with its annoying and unsafe consequences. One aspect of light pollution is **sky glow** or **uplight,** the result of too much light shining upward. It washes out the view of the night sky and any celestial bodies.

Town of Kennebunkport Outdoor Lighting Ordinance

Statement of Need and Purpose: Good outdoor lighting at night benefits everyone. It increases safety, enhances the town's nighttime character, and helps provide security. New lighting technologies have produced lights that are extremely powerful, and these types of lights may be improperly installed so that they create problems of excessive glare, light trespass, and higher energy use. Excessive glare can be annoying and may cause safety problems. Light trespass reduces everyone's privacy, and higher energy use results in increased costs for everyone. There is a need for a lighting ordinance that recognizes the benefits of outdoor lighting and provides clear guidelines for its installation so as to help maintain and complement the town's character. Appropriately regulated and properly installed, outdoor lighting will contribute to the safety and welfare of the residents of the town.

This ordinance is intended to reduce the problems created by improperly designed and installed outdoor lighting. It is intended to eliminate problems of glare, to minimize light trespass, and to help reduce the energy and financial costs of outdoor lighting by establishing regulations that limit the area that certain kinds of outdoor-lighting fixtures can illuminate and by limiting the total allowable illumination of lots located in the town of Kennebunkport.

The ordinance goes on to define terms such as fixture and glare and the installation requirements.

Electrical Elaboration

There are national and regional groups that address outdoor lighting. The International Dark-Sky Association (IDA) is a nonprofit, membership-based organization that educates its members and the public about effective nighttime lighting and the problem of light pollution. (The IDA can be reached by e-mail at ida@darksky.org.) Rensselaer Polytechnic Institute formed the Lighting Research Center (LRC) in 1988. It is now the world's largest university-based center for lighting research and education.

Other Concerns

Poorly planned and installed outdoor lighting can be both inefficient and unsafe. Excessive glare can make it difficult for pedestrians and drivers to see their way comfortably in the dark. This might not be a public problem in the privacy of your own yard, but what if you install lighting at the edge of your front yard near a public sidewalk? An overly bright lamppost at the end of a driveway in an otherwise light-subdued neighborhood could give a driver a momentary blind spot.

How do you know if a fixture is producing glare? If the light is beamed directly to your eye (if you can see the bright lamp from a distance), you've got glare. It's a result of the luminance from the light being noticeably greater than the surrounding luminance to which your eyes are trying to adapt. Ideally, you want the lamp to light the ground because that's where you're looking to walk. It does not need to light the sky (if it does, you're using too much light).

The lexicon of outdoor lighting continues with the term "light trespass," which refers to light that shines beyond the intended area for illumination. This includes lights that shine into bedroom windows next door, producing crabby, sleep-deprived neighbors. For that matter, the light also can be shining into your windows because it is misdirected or a poor choice of fixtures.

Finally, there's the matter of wasted energy. Well-designed and well-installed lighting will do the job with low-watt lamps because the light is properly directed. The energy lighting up the sky has to come from somewhere, and that somewhere usually is a power plant, which has better things to do with its electrons.

Drilling and Digging

A yard full of lights is a lot of fun, but digging the trenches for the cable isn't. Cable running underground has to meet certain material and burial restrictions. If you just want to limit the fixture locations to the side of your house, standard 14/2 or 12/2 cable running through the walls will do the job.

You have two choices for running cable underground:

➤ You can use underground feeder (UF) cable.

➤ You can run separate wires in plastic or metal conduit.

UF cable comes with an extra-tough insulation that's resistant to moisture, rot, and fungus. It also can be directly buried in the ground. It cannot, however, be embedded in concrete or cement.

Conduit is a raceway system to house electrical cable. Some local codes require that residential electrical systems be contained in some kind of raceway system instead of running exposed non-metallic cable. The types of conduit include …

➤ Electrical metallic tubing (EMT).

➤ Intermediate metallic conduit (IMC).

➤ Rigid metal conduit.

➤ Rigid PVC conduit.

EMT is the lightest and easiest conduit to bend and install, but it also is the easiest to damage. It is unlikely that your local code would allow this material to be buried in the ground because it would corrode too easily. IMC is thicker than EMT, is galvanized, and is acceptable for burial. Rigid metal conduit is the heaviest metal conduit available, is the most expensive, and requires threaded fittings to connect the sections.

Rigid PVC is nonmetallic and won't corrode when buried. Individual sections are heated and then bent when required, and they are joined with PVC fittings and glue

Bright Idea

You can try out different lighting plans by stringing small table lamps around your yard on extension cords. Use low-wattage lamps as you would in outdoor fixtures. This isn't perfect, but you'll get a general idea of what light levels to use and where to place the fixtures.

Positively Shocking

Always check your local code regarding using UF cable versus conduit. The NEC allows for UF cable as a buried conductor without physical protection, but a local code might require some form of conduit. The requirement might be based on local soil or climate conditions.

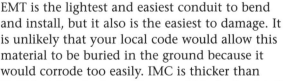

similar to PVC plumbing joints. PVC conduit can be attached only to plastic electrical boxes and always with a grounding conductor. Although Section 250-118 of the code allows for grounded metal conduit to act as a grounding conductor for metal boxes, local codes often call for a separate grounding conductor (the bare copper or green insulated wire in your cable) to be installed in the conduit. This is a good practice to follow regardless of your local rules.

Electrical Elaboration

We had a series of yard lights installed at our last house using PVC conduit. They were fine until some adolescent ne'er-do-wells decided to kick them over one night. They were re-paired, but the lawn-mowing crew knocked one over at least once a year. Plastic, it was clear, just couldn't support yard lights as well as metal conduit. I would never use it for this again.

Metal and Plastic

Metal conduit isn't flexible like electrical cable. Somehow you have to get it around corners and into tight spaces. An electrician will use a conduit bender to round the conduit sections (there are even small books written about how to bend it to various angles) and then use various fittings to connect them to boxes. A selection of elbow- and L-shaped fittings has eliminated the untrained homeowner's need for a conduit bender. It's a tedious world of pipe straps, setscrew couplings, and compression fittings that must be accurately installed. Any conduit for outdoor use must be properly sealed where sections are connected so that the wires, which must be rated for use in a wet location, stay dry.

PVC conduit isn't as strong as metal conduit, and it is bent by wrapping an electric heating element around an individual section. If you heat it up too much, you'll have to toss that section of conduit and try again. L-shaped fittings often can be used in lieu of bending.

You Can't Use Cable

Electric currents produce heat. A 12/2 cable might not feel warm to you, but heat is slowly dissipating from it as long as electricity is running along the conductors. If you encase this cable in conduit for anything other than very short distances, the heat buildup can be excessive and potentially dangerous.

So what do you do? You run separate conductors or wires instead. THNN/THWN wire is sold as individually insulated hot, neutral, and grounding conductors. THNN/THWN refers to the temperature rating of the insulation. THNN can withstand a temperature of 194°F and is the rating for most house wiring.

Dig It

Electrical wiring and outdoor elements don't mix very well. Any exposed runs, such as wiring running along the side of a garage, must be enclosed in conduit. You can't run UF cable either, even though it's rated for burial. The code likes to see your wire either buried or encased.

The depth at which you can run your wiring differs depending on the type of wiring you're using, the size of the circuit, and whether it's GFCI protected. The code requires that all outdoor receptacles be GFCI protected but not lighting or sprinkler controls. The following table lists the burial depths you must observe.

Bright Idea

Plan your landscaping and outdoor wiring at the same time so you can minimize the amount of digging you'll have to do. The best time to dig your trenches is after you've tilled the ground and broken it up. If you're not immediately ready to install the circuit, leave a narrow open trench and cover it with plywood to prevent anyone from falling in.

Depth Requirements for Common Residential Applications

Cable/Conduit	Circumstances	Depth
UF cable	GFCI-protected, 20-amp, 120-volt circuit	12 inches
Conduit	GFCI-protected, 20-amp, 120-volt circuit	12 inches
PVC conduit	Not encased in concrete	18 inches
UF cable/conduit for low-wattage use	Irrigation and landscape lighting controls not over 30 volts	6 inches

The Job So Far

At this point, you're looking at trenching your yard, calculating the load, running UF cable or conduit, and figuring out your lighting patterns. If you use conduit, you have to cut it to size with a hacksaw and use an array of fittings and connectors to install it. You can do this, but most people won't. Consider a compromise: Dig the ditches yourself or hire a digger to do them instead of your electrical contractor. Even if he or she puts an apprentice on the job, the hourly fee for ditch digging can start adding up. It doesn't take a lot of skill to swing a pick and dig with a shovel.

If you hire the electrical work out, have your electrician clearly determine where the trench should be dug, how wide it should be, and to what depth. Always cover a trench with planks, or at least place orange safety cones in front of it and connect them with yellow caution tape. You'll have to cover it if there's any chance someone could stumble into it at night.

Draw Up a Plan

A job you can do easily enough is drawing up the plans for your outdoor wiring. Determining the number of fixtures and devices, their locations, and the best approach to running your wiring needs careful consideration. You don't want to find yourself tunneling under any more sidewalks or sections of your driveway than necessary.

Section 210-52(e) calls for one outdoor receptacle in both the front and the back of a single-family residence and for each unit of a two-family dwelling where each unit is at grade level (in other words, a side-by-side duplex). These receptacles must be mounted no higher than six feet, six inches above grade level. These GFCI-protected receptacles can run on either a 15- or 20-amp circuit, but 20-amp is better. You want enough power to run power tools and plug-in yard tools without tripping the breaker.

Skip the Digging

If you can place your lighting fixtures and receptacles solely on the exterior walls of your house, you won't have to dig trenches or deal with conduit of any kind. Small spotlights or motion detectors attached to your siding can provide all the light you require for nighttime use. For a more decorative look, you can hang fixtures on your siding the same way you would hang sconces on your interior walls.

Outside receptacles usually are installed by cutting into the siding, although some homeowners install additional receptacles in gardens to power holiday lights or small fountains. You can install a receptacle on each side of your house, and this should give you plenty of options for temporary, plug-in lighting displays throughout the year. All outdoor receptacles must have weatherproof covers. If a plug is to be left in the receptacle unattended for an indefinite period of time (as opposed to temporary use such as an electric lawn mower), the code requires that the receptacle remain weatherproof while in use. This usually is done with a special recessed device and a special self-closing cover that is notched at the bottom so it can close over the power cord.

It's a lot easier to cut through your house's framing and fishing wires (using standard interior-grade cable) than to dig up your yard. It strictly depends on where you want your lighting located.

Light Options

There are two types of lights to choose from for outdoor lighting:

➤ Low-voltage (30 volts or less)

➤ Standard 120-volt permanent fixtures

Of the two, low-voltage lights certainly are easier to install, but keep in mind that you get what you pay for. The higher-end versions of low-voltage lights can be a pretty decent alternative to hard-wired, 120-volt systems. Cheap low-voltage lights are hardly worth installing, in my opinion.

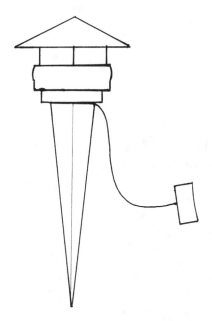

Typical low-voltage outdoor light.

How do you determine which system is best for you? Find an installation for each type (maybe a neighbor or a friend has outdoor lighting) and see how you react when it's turned on at night. Each system requires different installation procedures.

Low-Key Low Voltage

There's a range of outside low-voltage light systems. The simplest is available at most home-improvement or hardware stores, and it consists of a small number of fixtures, the cable connecting them, and a transformer to convert the 120-volt current to a 12-volt current appropriate for the system. The fixtures are one-piece affairs consisting of a lens cap, a lens, a reflector, a stake-shaped base, and cable. These are simple push-in affairs. The kits come with enough fixtures (typically six or eight) to line a sidewalk or perhaps one side of a driveway. More elaborate systems give you more versatility in designing a low-volt lighting system.

The cable used to connect low-voltage fixtures is rated for underground use and should be buried at least six inches deep in a narrow trench. The transformer must be plugged into a receptacle, which does not have to be GFCI protected. (If it's plugged into an outdoor receptacle, it automatically will be GFCI protected if it's installed to code.)

Higher-end systems, on the other hand, come with a series of direct burial transformers that allow for brighter lighting because they don't depend on a single, remote transformer that limits the wattage available at each fixture.

Low-voltage fixtures come with several lamp choices, depending on the model, including ...

➤ Halogen lamps.

➤ Small incandescent lamps.

➤ Five-watt fluorescent lamps.

Positively Shocking

Remember that, if you plug low-voltage lights into an exterior receptacle, it must have a weatherproof cover that can accommodate the plug and cord while closed. Your alternative is to mount the transformer inside your house and use an interior receptacle. The cable will have to run from the house through a short piece of conduit and then be buried.

Fiber Optics

A new low-voltage option is fiber-optic light kits, which range from basic plastic to long-lasting, cast-metal fixtures. Plastic fiber-optic cable runs between the fixtures. A lamp in the main fixture is the source of light for all the fixtures as its light travels along the fiber-optic cable. Some kits come with special 18-strand, side-glow cables whose entire length glows, making them ideal for marking sidewalks or driveways.

120-Volt Lights

Standard 120-volt light fixtures give you all kinds of options not available with low-volt systems. You can vary the intensity of the lamps with different-wattage bulbs and choose from a greater selection of fixture styles. Like any outdoor lighting, it should complement your landscaping and accent the architectural elements of your home. Fixtures in all price ranges and styles are available at lighting stores, and more limited selections are available at home-improvement centers.

Outside lighting provides security and shows off your yard and house. The 120-volt fixtures give off plenty of light in areas where you might have safety concerns. The use of timers and motion detectors also can enhance these fixtures. *You want the fixtures installed high enough so an intruder can't easily remove or loosen the bulbs!* It doesn't take much of a reach to remove a porch-light lamp, shake it to break the filament, and replace it, knowing that a homeowner might not replace it until sometime during the day. This can leave a front or rear door more vulnerable to an unwanted entry.

Fancy Fixtures

If you really want to go all out with your outdoor lighting, look into copper or bronze fixtures. Both will darken with age and take on a natural patina that some people find very attractive. These are not cheap, however! You should figure on spending around $100 per fixture (or more). Many styles also are very attractive additions to the day-time appearance of your garden.

The Least You Need to Know

➤ Outdoor lighting and receptacle locations have to be planned with as much scrutiny as your interior system.

➤ The most effective lighting avoids nighttime light pollution and unnecessary glare and shines on your property, not your neighbors'.

➤ Code requirements for outdoor wiring specify the type of conductors you can use, the conduit they can run in, and the depth at which they must be buried.

➤ Low-voltage lighting provides very subdued light for your yard, while 120-volt fixtures enable you to vary the light more by using different-wattage lamps.

Chapter 22

Electric Heat and Air Conditioning

In This Chapter

➤ Electric heat—pros and cons

➤ Zone heating

➤ Electric furnaces—big current draws

➤ Electric hot-water heaters

➤ Cooling off with air conditioning

➤ Wiring concerns

Home heating systems are one of our greatest civilizing forces, unless you live somewhere like Maui, where ghostly pale mainlander tourists really stick out in the winter months. Staying warm is one of our top goals when we live in cold, northern climates, and we have quite a few heating systems from which to choose: forced-air furnaces, zone heating, solar heat, and wood-burning stoves and fireplace inserts. Central heat uses some kind of furnace to heat air or a boiler to heat water. Furnaces are powered by natural gas, oil, electricity, or sometimes propane gas.

Gas generally is considered to be the most economical fuel for residential heating, but you'd never know it after reading comments and studies from its competitors. I'm not entering into this fray because there are too many variables to consider when it comes to heating cost and efficiency:

➤ The severity of your winter weather

➤ The age and condition of your current heating system

➤ How well your house is weatherized

➤ The temperature required for you to feel comfortable

➤ The cost of each power source

➤ The cost to install and maintain an individual heating system

If you live in southern California, it makes little sense to tear open your walls and install a gas furnace and its ducting when a few baseboard heaters might do the trick during the occasional cold spell. Freezing North Dakota residents, who probably never turn their furnaces off, most likely would find their electricity bills to be prohibitively expensive and consequently choose gas or oil.

Some neighborhoods might not have natural gas readily available and have few oil suppliers, so electricity might be the only option. New apartments and condominiums often have electric *zone heating* because it's cheaper to install than a furnace. This chapter deals with the ins and outs of electric heating and how it affects your electrical system.

Ask an Electrician

Zone heating is the opposite of a whole-house system in that it's composed of individually controlled heating units, such as baseboard heaters, that have their own thermostats. This enables you to selectively control the temperature of individual rooms or **zones.** The closest you can get to zone heating with a furnace is to shut down the ducts leading to individual rooms.

Electric Furnaces

To hear an electric utility or vendor tell it, you can't get a better source of central heating than an electric furnace. Utilities, of course, have a vested interest in electrifying everything in our lives, from opening a garage door to peeling a potato. Nevertheless, an electric furnace offers some advantages over other types of heat:

➤ Safety (There are no combustible fuels to deal with or have piped into your house with flameless electricity.)

➤ Design flexibility (No pipes, storage tanks, or vents are required.)

➤ No removal of humidity or oxygen from the air (Fossil fuels consume oxygen during combustion.)

➤ Few moving parts for lower maintenance

➤ Near 100 percent efficiency because there are no fuels to burn, toxic byproducts to remove, or possible leakage of fumes

An electric furnace is a fairly simple affair. A fan blows air across a heat exchanger or resistor, a wire coil heated by electricity, and into the ductwork. No individual circuit feeding the furnace can be larger than 60 amps. Installing an electric furnace requires coordinating the furnace installer with whoever is doing the electrical work.

Electrical Elaboration

Oil heat apparently isn't as dead as its competitors would like you to believe. Oil dealers report 130,000 conversions from electric heat to oil heat over the past five years. They further indicate that electric utility companies push electric heat pumps, which are cheap for builders to install for air conditioning/heating systems, but that the pumps are inefficient during really cold weather. The Oil Heat Manufacturers Association claims relatively short payback periods for conversions as well as overall savings during the life of the system. Gas and electric utilities have their own figures, of course, challenging those of the oil dealers.

Seal Those Ducts

Furnaces have two sets of ducts or tubes: One carries heated air to registers in different rooms; the other pulls air back to the furnace through the cold-air returns. Central air-conditioning systems share the same ductwork to distribute cold air throughout the house. Leaks in your ductwork not only can cost you money in lost heat and cool air, they also decrease the efficiency of your system.

You should check your ductwork at the beginning of the heating and cooling seasons or if you don't feel enough air flowing out of an individual register. A big problem is disconnected ducts, particularly with flexible ducting (as opposed to rigid sheet-metal ducting), in unfinished areas of your house such as crawl spaces.

Electrical Elaboration

We purchased one of six new houses in a small cul-de-sac in the fall of 1994. All of these houses have accessible crawl spaces instead of basements. When the cold weather hit, half of us discovered that sections of our furnace ducting weren't properly connected. Our neighbor made all kinds of subterranean creatures really happy as a slipped duct heated his crawl space to a toasty warm temperature. The heating contractor came back and made the necessary repairs (sort of), but he was never hired by the building contractor again.

Standard gray duct tape has been used for years to seal duct joints. Oddly, a recent study suggests that duct tape is good for just about everything *but* sealing heating ducts, a finding disputed by manufacturers. The argument against depending on duct tape is that the adhesive loses its effectiveness over time, and the ducts eventually will leak. Duct mastic, however, is a soft, putty-like material that seals joints and leaks very effectively. Applying duct mastic and backing tape when needed is a nuisance, and I've personally never seen a furnace installer use it (that doesn't mean that some don't). Realistically, applying several extra layers of duct tape is the most a home-owner can be expected to do unless the duct is very accessible.

It's important that ducts be properly supported so they don't drop, sag, and develop breaks in their connections. Metal ducts can be supported with plumber's tape, which is a thin, perforated, metal tape for holding pipes and drains in place below floor joist. Flexible ducts, used regularly in new housing, should be supported every five feet and within six inches of any connection.

Zone In

The most common heating system in detached houses is a central system employing some type of furnace or boiler. Apartment buildings used to be heated by steam or hot-water heat using large boilers, but multiple dwellings are being built increasingly with zone heating. The advantage for the owner, besides lower construction costs, is that individual heaters shift the cost of the heat to the renter.

Positively Shocking

Zone heaters are not entirely maintenance-free, but they are close to it. Both resistance and radiant heaters need occasional dusting, but it's more critical with resistance models. Dust and lint can build up in and around the fan and heat conductors, cutting down on the unit's efficiency. In extreme cases, a fire could start. Use the plastic tool attachment that came with your vacuum cleaner to clean these.

In your home, you might install zone heating in a new addition if it's impractical to extend your existing central heating system. You also might install individual heaters in rooms that already are serviced by heat registers or radiators but that still don't get comfortably warm. Bathrooms in older homes are a prime example. Zone heating enables you to control your energy use by keeping less-frequently used rooms cooler, and maintenance costs are lower than those of a central furnace.

One drawback to individual heaters is having to manage one thermostat per zone instead of a single thermostat for the entire house. The simplest thermostat is an "On/Off" switch that's either attached to the heater itself or wall-mounted and that's gradated for temperature. More expensive thermostats are programmable and enable you to set each room on its own schedule according to use. Some utilities offer incentives for all-electric homes with off-peak, discounted rates. I've also read of others that offered multiple-year, locked-in rates.

Hooking Them Up

Both blower-style resistance heaters and baseboard types are made for 120- or 240-volt ratings. The wire gauge will depend on the number of heaters (and their wattage) connected to a dedicated circuit. The code states that no heating circuit shall be larger than 30 amps when feeding individual heaters. The wire will be connected to the heater's thermostat, which in turn is wired to the junction box on the heater.

Baseboard heaters can be configured in any number of ways depending on your heating needs. They can be installed as single units of varying lengths, or two or more units can be joined together. Metal elbows are available for fitting the heaters around corners.

Heaters with Receptacles

Some baseboard heaters conveniently come with receptacles installed. Great, you think, two devices on the same circuit. Sorry, the code (Section 210-52) requires that the receptacles be wired to a separate circuit, but they can be counted as one of the required room receptacles.

Neither existing receptacles nor new ones can be installed above a heated section of a baseboard heater. They can be installed above *spacer sections* of a baseboard, which resemble the rest of the unit but do not emit any heat. They are available for just this type of installation, in which a receptacle's location near the baseboard would cause a code violation.

Going Portable

Portable space heaters are inexpensive and are a great way to tax your electrical system. Used properly, a space heater can warm up a cold room that your furnace seems to forget about, or it can make an unheated basement workroom or hobby room more user friendly. Because space heaters produce heat directly, there is no heat lost in ductwork, as occurs to some extent with furnaces. Portable, plug-in heaters come in two styles:

➤ Radiant
➤ Resistance

A radiant heater is similar to a baseboard heater or a radiator. (Some are even shaped like a radiator.) They heat objects (walls, floors, people) as their own surface heats. Most resistance heaters use a fan to blow air across a hot coil or another conductor. Either heater can do the job for temporary heat, but you must follow the rules for safe usage:

➤ Be sure the circuit in the room can support the load imposed by the heater. (A 1,500-watt heater, for example, must be on a 20-amp circuit.)
➤ Never use a portable heater in a bathroom or around water.

➤ Plug the heater directly into a receptacle, not into an extension cord.

➤ Never leave a portable heater running unattended.

➤ Keep the heater away from anything combustible (paper, clothing, and so on).

Bright Idea

If you must use a portable heater, consider the radiant type. There's a smaller chance that the unit itself will ignite anything near it because it won't have a red-hot conductor providing the heat source. You'll still have to confirm that the branch circuit can support the load, however.

Radiant Heating

Technically speaking, radiant heat warms with long-wave electromagnetic radiation, distributing rays of heat fairly evenly in a 160-degree arc. By affecting objects instead of the air, radiant heat at a lower temperature can warm a room faster than other types of heat at a higher temperature. Radiant heat systems are installed either in a floor, from which the heat will naturally rise, or in a ceiling, an intuitively bad location given the aforementioned nature of rising heat.

Radiant heat has never been a huge seller in American homes, even though it makes a lot of sense in theory. A uniformly warm floor with rising heat should be more comfortable than trying to distribute heat through furnace registers or radiators. Systems are not just limited to concrete floors, and some are available for retrofitting wood floors.

Some of the systems available include ...

➤ Cables embedded in wet concrete floors.

➤ Radiant heat foil for wood-framed floors. (For ceiling applications, foil is stapled to the face of the ceiling joist and then covered with drywall.)

➤ Wafer-thin radiant floor mats installed under tile, carpet, wood floors, and so on with cement underlayments such as Wonder Board.

As a source of heat, radiant floor and ceiling systems require dedicated circuits, usually 240 volts. The heat is controlled by thermostats, timers, or simple "On/Off" switches.

Heat Pumps

Heat pumps once were touted as an economical response to rising energy costs. They work by removing heat from outdoor air and moving it inside your house in the winter months. In the summer, the process reverses, and the pump removes hot air, leaving your interior cooler. Heat-pump detractors claim that they do not work well or efficiently in harsher climates.

The general response to heat pumps hasn't exactly been overwhelming. You'd think that in the Northwest, with its generally mild winters and nominally environmentally responsive citizens, we would have more homes using heat pumps. They're not a big seller, however. Heat pumps require a dedicated circuit.

Whole-House Ventilation

The old, beautiful homes that many people pine for are a pretty leaky lot. Insulation was not ubiquitous when these homes were built, as it is today, and insulated windows were barely existent (although I have seen one early version dating back to the 1930s). We're building tighter houses, which means lower heating bills, but now we need to build in mechanical ventilation systems for fresh air. We have to vent out the various gases, carbon monoxide, and moisture produced by our homes and, well, us. Paints, cleaning products, carpets, and vinyl continue to cure for various lengths of time after they're applied or installed. Cooking produces steam and odors, and every time we breathe we expel moisture. It all has to go somewhere, and outside is a good place for it.

Positively Shocking

Heat pumps should be repaired by technicians who are certified to do so. Not all heating and cooling personnel are qualified to do these repairs. Certification is granted after a technician has been trained in heat-pump repair and has passed a test given by the Refrigeration Service Engineers Society (RSES).

A ventilation system in a new house typically has a ceiling fan installed in an out-of-the-way area such as a laundry room. The fan is attached to a timer that you can set to go off at regular intervals to ensure that air exchanges happen throughout the day. Some systems even have a heat-recovery capability that retains heat from the expelled air.

Ventilation systems do not require dedicated circuits or GFCI protection. They can run on a 15-amp branch circuit unless otherwise noted in their installation instructions.

Air Conditioning

A good percentage of the country would be almost uninhabitable by modern standards during the summer months if it weren't for air conditioning. Central air-conditioning shares the same ductwork as central heating. Room air conditioners, on the other hand, are installed either in a window or through an opening in the outside wall and are plugged into a receptacle. Individual room air conditioners are strictly for zone cooling. If you want your entire house cooled, it's usually cheaper and quieter to run a central system.

Central air conditioning gets its own 240-volt dedicated circuit. The nameplate on the side of the air-conditioning unit spells out all the electrical requirements. You should discuss these with your heating and cooling contractor so the circuit wiring will be completed before the unit is installed.

Individual room air conditioners come in both 120- and 240-volt ratings. The individual unit will determine the circuit requirements. The code, however, does call for the following:

➤ The unit must be grounded.

➤ It must be connected to a receptacle using a cord and plug.

➤ The unit cannot draw more than 50 percent of the branch circuit's ampacity if it shares the circuit with other devices. On a dedicated circuit, it cannot exceed 80 percent of the circuit's ampacity.

➤ The air conditioner cannot have a rating higher than 40 amps.

If the unit you want to install is rated at 240 volts, you'll need a special 240-volt receptacle.

Air-Conditioning Alternatives

Air conditioning was once a luxury. Industry figures indicate, however, that by the late 1980s, 64 percent of American homes had air conditioners. In hot climates, air conditioning can really run up your electric bills in the summer months. Rather than depending solely on mechanical cooling, however, we can take a few tips from the days before such systems were available and let nature lend a hand.

Three considerations will help keep you and your home cool in the heat of summer:

➤ Shade from landscaping

➤ Building design

➤ Mechanical ventilation

These will not eliminate humidity like air conditioning does, but they can create an environment in which your cooling system won't have to work as hard to keep your house comfortable.

Shade

Trees, shrubs, long overhangs, awnings, and window coverings all provide shade that will decrease the heating effect of sunlight on your home. Trees and shrubs, especially large deciduous trees, can shade entire walls and sections of your roof. In the winter when they shed their leaves, they allow sunlight in during the time of year you want its radiant energy. Coupled with shrubs, properly placed trees also can help cut down on winter winds. Shrubs don't get as large as trees, but they grow faster and are less costly to purchase.

Awnings can be fixed or the roll-up variety, the latter of which can be made from cloth or metal. Prices and installation costs vary, so be sure the expense is justified by the results. In moderately hot climates, awnings might not be economical. Other window treatments include …

➤ Heat-reflecting coatings that attach to the glass.

➤ Window shades.

➤ Blinds, drapes, and shutters.

New Construction

If you're starting out with a fresh, clean lot and a house plan, you can design all kinds of energy-saving features into your new home. Your architect can assist you in siting your house for the best use of winter sunlight. A landscape architect or designer can advise on plant and tree selections for the most effective summer cooling. Long overhangs will keep your second floor cool and will protect your siding from rain. None of these factors will change your electrical load calculations, but they can cut your electricity bill!

Positively Shocking

Don't plant your shrubs or other plant life too close to your siding or windows. Plants can prevent siding from drying out during wet weather and can promote the growth of fungus and, in the worst cases, rot. A three-foot distance or so between your house and the plants will prevent them from holding in too much moisture and will allow the house to dry.

Hot-Water Heaters, Big and Small

Unless you're living in Iceland, the land of thermal springs, your water most likely is heated in a large (usually 40 gallons or more) tank using gas, oil, or electricity. Once again, the competing energy suppliers all pipe in with opinions as to which is the most efficient and economical hot-water tank. The one similarity they do share is that, in some markets, the tanks can be either purchased or leased, depending on the programs offered by your local utilities.

Electricity advocates cite the following advantages:

➤ There's no pilot-light or electronic ignition, leading to more dependability and fewer repairs.

➤ There's never any danger from carbon-monoxide fumes.

➤ Installation is easy because an electric water heater doesn't have to be vented.

➤ You never have to depend on an oil delivery.

One factor that electrophiles don't mention, however, is the loss of hot water if the power goes out for any extended length of time. Furnaces depend on electricity to run the fans, so a power outage affects them all equally. A gas or oil hot-water tank, however, doesn't depend on electricity unless it has some kind of an electronic ignition. A more common pilot light will, like a certain drum-playing lagomorph for a certain battery manufacturer, just keep going and going and going.

Electrical Elaboration

There's one type of furnace that doesn't depend on a fan to distribute warm air: a gravity furnace. These are ancient and rarely seen today. The heat simply rises up and through a limited system of ducts (which means good luck if your bedroom is on the third floor of the house). These furnaces might be interesting from a historical standpoint, but they are not practical.

Mixing Water and Electricity

Electric water heaters have varying wattage ratings depending on their capacity in gallons and their speed of recovery (how fast they can heat incoming water). Wattage can run from 1,500 to 5,500, depending on the water heater. The unit will require a 240-volt circuit, most likely 30 amps (the minimum required by code) using 10-gauge wire. This, too, will vary depending on the tank you choose.

Hot Water on Demand

Tankless (or "instantaneous") water heaters heat water on demand rather than storing hot water in a tank. (These were mentioned in Chapter 20, "Bathroom Wiring.") They are common in Europe, where space and energy savings are more pressing concerns than in the United States.

A tankless water heater ...

➤ Never runs out of hot water.

➤ Is considered an energy-saving appliance in many areas.

➤ Is located at the point of use, so it does not require a single large space for installation.

➤ Requires a minimum flow rate to activate.

➤ Will not heat a high flow rate of water as comfortably as a lower rate.

➤ Requires its own circuit unless it's gas operated (running a gas pipe to each unit).

Because these are point-of-use appliances, you have to install more than one of them to supply hot water to all your plumbing fixtures. This is hardly a likely scenario in a modern household, at least on this side of the Atlantic. If you believe you'll get some energy savings out of them, you'll have to install a dedicated circuit for each unit. Keep in mind that an insulated water tank maintained at 120°F (often a maximum setting when regulated by local laws) in an area of low fuel or electricity costs also can be energy-efficient.

The Least You Need to Know

➤ There are too many usage and cost variables to all-inclusively state that one type of central heating system is best for you.

➤ An electric furnace is clean and simple to operate, but check local rates and utility incentives before deciding on one power source over another.

➤ Zone heating is an efficient way to warm up a cold room or two without the mess of extending your furnace ducting.

➤ Portable heaters are convenient, but never treat them as casually as a built-in heater.

➤ Air-conditioning and heating costs can be cut down, and your house can be made more comfortable, by some creative landscaping and window coverings.

Part 5
Refinements

Electricity provides us with all kinds of opportunities to pamper ourselves with toys, gadgets, and electronic indulgences that Edison could only have dreamed about. Automatic window blinds? No problem. Security systems that sound like barking dogs? One fake German shepherd coming right up. And don't forget a phone system that's more sophisticated than those of the governments of some small countries.

It's easy to take a lighthearted look at home automation, but some devices can serve worthwhile purposes. You might hear a knock at the door, but door chimes are better and more reliable. Smoke and fire alarms can save lives. Few of us want to be without high-speed Internet access, which means running appropriate wire to computers (and wiring computers to each other).

The other side of this electrical wonderland is awareness of consumption and knowing when enough is enough. Do we just keep grabbing every watt at any price? Or should we consider some alternatives to decreasing our power usage? The Department of Energy and your local utility have an abundance of free information that will advise you how to achieve energy conservation without putting a noticeable dent in your lifestyle. Ideally, you should install an electrical system that enables you to take full advantage of modern conveniences, but you should have the wisdom not to abuse it.

Workshops, Offices, and Generators

Kitchens and bathrooms are two types of rooms that often need electrical upgrading. Our usage is more intense in these rooms than, say, in the living room where we read and socialize. Our only electrical demands in the living room are for reading lights and receptacles for sound systems or maybe the TV.

Hobbyists, woodworkers, and people with home offices have broader needs, however. A woodworker needs plenty of power to run saws, drills, and routers as well as sufficient lighting to work safely. Someone who works with electronics also needs a well-lit workbench and sufficient ventilation to remove fumes from soldering. Home office workers have plenty of equipment (computers, printers, copiers, and so on) that would benefit from dedicated circuits. For people who regularly lose their electrical power in windstorms, a generator certainly is worth consideration.

All of these projects require running electrical cable and installing devices. The results will be convenience, safety, and much less risk of overloading an existing circuit. Wiring a workshop or an office is no more difficult than wiring any other room of your house. Generators require a qualified electrician, however, if they are to be directly connected to your electrical system.

Workshops

The term "workshop" means different things to different people. To many, it implies a woodworking area replete with an assortment of power tools: saws, planers, sanders, a drill press, and routers. To others, it's a place to create stained-glass windows, to throw a pot or two and then fire them in an electric kiln, or to work on jewelry designs. All of these workshop types need ...

➤ An adequate number of receptacles for plugging in electric tools and gadgets.

➤ Sufficient power, usually on a dedicated circuit to run these same tools and gadgets.

➤ Plenty of light, especially for close work.

➤ In many cases, some form of ventilation.

The more power your hobby or work demands, the more demanding your workroom's electrical needs.

Bright Idea

While you're pulling wires for receptacles and lights, consider installing cables for a telephone and an intercom speaker.

Woodshop Details

There is no shortage of power tools available to the amateur and talented woodworker, but you can only run one or two of them at a time. For this and other reasons, a workshop does not require the same number of receptacles as the finished parts of your house. Because most workshops are located in basements or garages, both finished and unfinished, the following rules do apply:

➤ A dedicated circuit should be used for any tool or device that will draw more than 50 percent of the circuit's ampacity.

➤ Any remaining receptacles also should be on one or more dedicated 20-amp circuits to run smaller tools.

➤ Your receptacles must be GFCI protected unless the receptacle is not readily accessible or is used exclusively for a fixed-in-place appliance such as a freezer.

➤ Your work lights should be part of a separate branch circuit from your receptacles.

You don't want the lights to be on the same circuit as your receptacles because an accidental overload could trip the breaker. You'd find yourself standing in the dark, which isn't such a good idea in a workshop. Can you assume that you'll be the only person working in the area? If two people are running power tools, you could easily overload a single circuit. For this reason, it might be best to run two circuits, as you would in a kitchen.

There are table saws and then there are, well, bigger table saws. Sears' Craftsman line of power tools, for example, features a 10-inch Benchtop Table Saw that runs handily on a standard 120-volt circuit. The Professional Table Saw is quite another story and features a 120/240-volt motor. You don't want to find out after you've run your cable that the circuit you've installed isn't adequate to do the job. You need to know in advance the type and size of tools you'll be using.

Raceways or Cable?

If your workshop is in an unfinished area (one with exposed framing without drywall or plaster), any accessible electrical cable must be protected with a wall covering or be contained in a raceway of some kind. You can't have a cable running through exposed wall studs where it can be damaged. Consult your local code regarding cable running through exposed ceiling joist. This is not usually viewed as being accessible to damage because it's out of reach from most activities.

There are a lot of rules regarding conduit or raceway installations. These rules cover the following:

➤ Bends and offsets

➤ The cutting and fitting of conduit sections

➤ How to fasten the conduit to walls and ceilings

➤ The correct size of raceway for the number of conductors being run (taking into account their gauge and type)

➤ Transitions between exposed cable and conduit

Unless you're installing a system like Wiremold (described in detail later in this chapter), I recommend that you leave any extensive raceway installations to an electrician.

Safety Power Controls

You have two major protective devices in your workshop area:

➤ GFCI receptacles

➤ Circuit breakers

Bright Idea

Take a small notebook or an electronic organizer with you and browse through a tool store. Record the wattage and voltage requirements of all the tools you think you'll eventually be stocking in your workshop. (Try to be realistic.) Use these totals to determine the circuitry in your workshop so you only have to wire it once.

Positively Shocking

Wearing thick work gloves while using some power tools is not a recommended practice. It's too easy for the fingers of the gloves to get caught in moving parts such as saw blades or drill bits.

It would be useful to have a switch-controlled receptacle if you have to quickly turn off a power tool with a broken trigger and cannot conveniently pull out the plug. All the receptacles on a workroom circuit can be switch controlled. A switch also can act as a safety device if your children ever get into your workroom unsupervised.

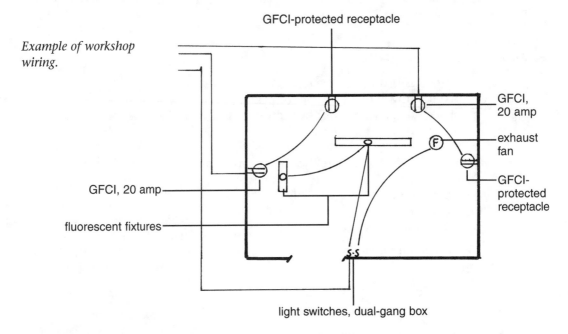

Example of workshop wiring.

Lighting Your Workroom

Fluorescent fixtures are a good choice for basement workrooms. A couple of two-lamp fixtures will provide plenty of light and will eliminate most shadows when working over a bench or a standing power tool. Some work, such as soldering electronic circuits or leaded-glass joints, requires a more focused light. This need can easily be remedied with a plug-in, swing-style incandescent light that will utilize one of your GFCI receptacles.

Fluorescent fixtures are more problematic in unheated garages in cold climates. The farther the temperature drops below 50°F, the more difficulty fluorescent fixtures have in starting, and they might not start at all. If you do install fluorescent lighting in an unheated area, be sure to use fixtures with cold-start ballasts.

A Breath of Fresh Air

A workroom can always use some kind of ventilation from either a window or a fan. This is especially important if your work involves soldering, paints, or other finishes. A ceiling or wall fan can be installed in the same manner as a bathroom exhaust fan, and it can be run off the lighting circuit.

Home Offices

If you believe various futurists, many of us will soon be working out of home offices for at least part of our working hours. Many people already do work at home for either a full-time job or a home-based business. Our homes are full of possible office sites including …

➤ Spare bedrooms.

➤ Attics and basements.

➤ Detached garages.

➤ Spaces under staircases.

The location of your office might affect your wiring choices. The sophistication of your equipment and the number of phones, computers, and computer peripherals you have will determine the number of circuits and receptacles needed. If there's one rule to remember about a computerized, online office it's this: Even your most top-of-the-line equipment today will be outdated sooner than you think. Phone and cable companies are all clamoring to hook you up to the Internet, and computers are getting faster, cheaper, and loaded with more memory even as you read this book. You can't plan for every wiring contingency, but you don't need to be caught blindsided, either.

Beyond Manual Typewriters

Once upon a time in home offices long ago, people had one rotary-dial phone, one manual adding machine, one manual typewriter, and a desk lamp. It can be tough enough explaining to your kids such a Spartan existence, but try explaining a typewriter. It was a big deal when electric typewriters became household commodities.

Typical home-office equipment these days includes the following:

➤ One or more computers

➤ Telephone/fax

➤ Printer

➤ Zip drive

➤ Scanner

➤ Copier

You could try to get by with most of this equipment running on an existing branch circuit, but you're taking the chance that a load somewhere down (or up) the line might trip the breaker. A sudden loss of power also can mean the loss of computer data or the interruption of a fax. An office should get its own dedicated circuit(s). A copier, depending on its wattage, definitely should get a dedicated circuit.

You should have a minimum of one dedicated, 120-volt, 20-amp circuit in your office. You might consider a second circuit if you're running more than one computer and its peripherals. You don't want both of them going down at the same time if a breaker trips. If your office is located on the third floor of your house or in a detached, remodeled garage (anywhere far from the main circuit panel), you should consider installing a subpanel.

Voice and Data Coming Through

Through the Bush administration, the White House, home of the president of the most techno-savvy country in the world, still used telephone operators working a switchboard. Say what you will about Bill Clinton (unfortunately, there's plenty to say), he did bring the White House phone system up to modern standards, presumably to the chagrin of the no-longer-needed operators.

It's tough to run a home office with just one phone line given the need for Internet access and sending or receiving faxes. You might have multiple phones on one line and need to wire them properly, and data transmission can be annoyingly slow with the wrong wiring or service access.

There are a few basic rules to follow:

➤ Follow a *star wiring pattern* with your phones.

➤ If three lines are too expensive, you can get by with two if your system and equipment are flexible.

➤ Install a run of conduit for future use from your office to the basement or garage where your phone lines come into the house.

➤ Buy the highest-rated wiring you can.

Ask an Electrician

A **star wiring pattern** means each phone jack is wired directly to a central wiring point. This is different from a **daisy-chain pattern,** in which phones are linked in a series, one to the next. A star pattern allows for easier diagnosis and repair of individual phone problems.

Chapter 24, "Your Own Hi-Tech Revolution," deals with the specifics of the phone/data/cable conundrum, but from a wiring standpoint, you want to make your office as up-to-date as your budget will allow. At the same time, you want to be prepared for the inevitable changes and breakthroughs in technology that seem to be introduced on a monthly basis.

Consider wiring phone, coaxial, and fiber-optic cable all at the same time, even if you don't have any immediate use for each service. Consult with several telephone service providers before settling on one phone system. Phones are programmable now, and this gives you a lot of options when setting up your system.

Conduit for Future Options

A one-inch-wide section of metal or plastic conduit running between your office and a central wiring point for incoming voice and data cable will allow for easy future upgrades to your office. It's going to be some time before we all go wireless (and the cost will be anyone's guess), so figure on wired connections for now and the immediate future. You don't want to have to go tearing into your walls and pulling off trim to speed up your phone and data connections. An empty piece of conduit running to a convenient endpoint in your office will enable you to take advantage of new technologies.

Instead of fishing your future cable through the conduit, you can take the easy way out. After installing the conduit, run a heavy piece of string through it and leave it in. You can use it to pull cable through the conduit at a later date. The procedure is simple, and I have to thank the folks at the television show *Hometime* for the tip. Simply follow these steps:

Bright Idea

It might be tricky running conduit across wall studs without tearing into the plaster or drywall too much. You can always run it up vertically from the basement and let it end in a closet. This at least gives you a starting point for future wiring.

1. Attach a small piece of cardboard or wadded paper to one end of the string.

2. Insert this end into one end of the conduit.

3. Attach the hose of a vacuum cleaner to the other end of the conduit.

4. When you switch on the vacuum, the string will be pulled through the conduit.

5. Give yourself an extra foot or so of string on each end, and tape it to the conduit.

6. When you need to install wire in the conduit, simply attach to one end of the string and pull on the other.

How Fast Can You Go?

The information revolution has given us far more readily available data than any of us can handle. The more information, the faster we need to browse and say "yea" or "nay" to it. This is where your choice of cable is important. Category 3 cable is fine for most data transfers at 10 to 16 megabits/second, but category 5, with its better insulation and twists per foot (see Chapter 24), can transfer data at 100 megabits/second. The future will demand more speed, not less. For your work purposes, this might not be an issue, but you should keep it in mind when you wire your office.

Positively Shocking

Remember, your electrical cable has an electromagnetic field that can interfere with phone, data, and cable TV wiring. You should not run them together in the same holes in your studs and floor joist, nor should they be run together in pipes or duct-work. Run them on opposite sides of the stud and joist spaces or otherwise 12 inches apart.

The Need to Suppress

I've already discussed surge suppressors (see Chapter 9, "Extension Cords and Multiple Strips"), but it bears repeating here. Your office will likely have thousands of dollars worth of equipment. A suppressor is an inexpensive investment whether it's a desktop device or is located in the service panel. Be sure that all your electronic equipment and phones are protected. Remember that surge suppressors do not work without a grounding conductor, so this must be present in your wiring.

An Outside Job: Raceways

A raceway is a metal or nonmetallic protective container for wires or cables that often, but not always, runs along a wall or ceiling surface. Any metal conduit is one form of a raceway, but there are others such as …

➤ PVC conduit.

➤ Underfloor raceways.

➤ Surface raceways.

➤ Wireways.

Electrical Elaboration

Just because you're using conduit or a raceway doesn't mean your wire or cable is suddenly invincible and cannot be damaged. Like water pipes, conduit can still be damaged if it's not properly installed and protected. I watched one worker drill into a wall and through a cold-water pipe while installing a bathroom mirror. You'd be surprised how fast a couple of rooms can flood before a job superintendent manages to find a master shutoff for a four-story building. You'll get less-visible problems if your conduit is ruptured, but it could be just as bad in the long run.

In a residence, you're most likely to use a raceway when it's impractical, too costly, or too much work to run electrical cable through the walls or floors. In some cases (such as installing a device in a basement laundry room with concrete walls), it's essentially impossible to run a cable in the wall. In other cases, a raceway in a closet or another out-of-the-way area won't detract from the room's appearance.

Conduit can be a little intimidating to work with, but one form of consumer-friendly raceway is made by, and named, Wiremold.

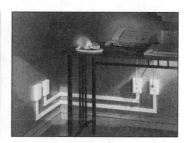

Wiremold raceways.

It's a Wiremold World

Wiremold is a complete system that includes the raceway itself, clips, couplings, boxes, and cover plates. The Wiremold Company (www.wiremold.com or 1-800-621-0049) manufactures a number of easy-to-use products including ...

➤ Metal raceway

➤ Nonmetallic raceway

➤ CordMate I and II

➤ Corduct

➤ Plugmold

Both metal and nonmetallic raceways are installed in a similar manner. A starter box (which acts as the electrical box for a switch, receptacle, or fixture) is surface-mounted in the desired location. The metal or nonmetallic raceway is cut to size with a hacksaw and then is mounted on the wall or ceiling with clips or couplings supplied with the raceway. Individual wires are then installed in the raceway.

With metal Wiremold, the wires are pulled through; with the nonmetallic version, the raceway has a snap-on cover. When the cover is removed, the raceway is open for easy wiring. Once the wires are in, any fittings, devices, and cover plates are installed. Both the metal and nonmetallic raceways can be painted. (The nonmetallic also can be stained.)

CordMate I and II are protective raceways for lamp cords, extension cords, and telephone wires. They conceal the cords for a neater appearance and to prevent anyone from tripping over them. CordMate is a plastic raceway that is cut to size and attached to a wall with an adhesive backing. After the cord or wire is laid in, the CordMate is snapped closed.

Corduct works in a similar manner to CordMate, except it adheres to your floor. Plugmold is basically a power strip of up to 10 receptacles that is mounted to a wall bracket and then plugged into an existing receptacle.

A raceway running along a wall or a ceiling is a compromise approach in a finished room. You're choosing to update or improve your electrical service without tearing into the walls. The advantage, of course, is the elimination of plaster or drywall repair and repainting the room. The drawback is having to look at the raceway. No matter how you paint it, it will stick out. Raceways are a common solution in older apartment buildings when an owner is reluctant to disrupt tenants by poking holes in their walls and ceilings. Raceways can be installed with little muss or fuss. In an unfinished basement, the presence of a raceway isn't an aesthetic issue at all.

Positively Shocking

A raceway is not meant to cover up old, fraying lamp cords. These should be replaced before the cords are routed through the raceway. Otherwise, you run the risk of a hidden electrical fire.

You also can view a raceway as a temporary solution to wiring one or two rooms in which additional receptacles or fixtures are needed temporarily until you do a more thorough remodeling in the future. Sometimes the future is farther away than you plan. In the meantime, you have the convenience of up-to-date wiring.

Generators

Unless you're living in what real estate agents euphemistically refer to as a "rustic location," you probably don't think about portable generators very much. In areas that are not served by reliable electrical service, a generator, if properly installed, is a good addition to any household.

A generator produces AC voltage rated in watts, which is compatible with your electrical system. You can't, however, simply go down to your local home-improvement store or cruise the Internet, order a generator, haul it home, and consider your task finished. You have to know how to run it and, more important, how to hook it up to the loads you want it to run. An incorrectly connected generator can endanger utility workers repairing power lines as well as you and your family. A generator can be cheap insurance during an extended power outage, but you need to start with the right information.

Generators can be used for many things including …

➤ Standby emergency use for homeowners.

➤ Camping.

➤ Construction rentals.

Standby use is our main concern in this chapter.

Electrical Elaboration

A new series of mini-turbine, gas-powered generators is being developed for single-building use. These units aren't all that much bigger than a refrigerator and can produce enough power to run a small factory, fast-food restaurant, or several homes. Some suggest that these generators could offer real competition to electrical utilities and eventually will be small enough to comfortably supply electricity for single houses.

The Mechanics of a Generator

Think of a generator as a small version of your utility's power plant, but instead of a gas- or water-powered turbine spinning the magnets, a gasoline- or diesel-powered engine is doing the job. The generator's circuits (either 120-volt or 120/240-volt) then distribute the currents to your loads through some kind of power-cord connection. The generator will keep running as long as it has fuel in its tank.

The power output of a generator can be described in two ways:

➤ The maximum power is its absolute limit. A generator cannot sustain this level indefinitely.

➤ Rated power, which is 90 percent of the maximum-power figure, can be produced for much longer periods of time.

You can run a generator at its maximum power output, but only for a half-hour or so.

Makita generator.

A generator has to deal with loads (all the devices you want it to run). If the power goes out and you need to use the generator, you want to get the best use out of it. This means you'll be running some necessary lights, maybe the refrigerator, and a radio or television for news. It doesn't mean you should plug in hair dryers and portable heaters because they will quickly eat up the capacity of most generators. Loads are defined as one of two types:

➤ Resistive
➤ Reactive

The loads you run will determine the generator you purchase.

Resistive and Reactive Loads

A resistive load usually is anything that does not involve an electric motor. An incandescent light fixture and a television are resistive loads. A reactive load involves an electric motor. Motors need additional power to get them going initially before they reach running speed. A furnace fan and a refrigerator are reactive loads.

Bright Idea

Be sure to test your generator before you actually need it. Do a test run, and be sure at least one other responsible person in your household (or a neighbor) knows how to start it. You want to discover any problems before an emergency occurs.

Your generator's supplier should have a load-wattage-estimate chart to help you figure out which generator will suit your needs. Ultimately, your generator must power all your resistive loads and must handle the startup wattage of any reactive loads. You'll want a generator at least one size greater (in terms of wattage) than your total load.

Electrical Elaboration

Some solar-powered generators are available, but they're not cheap, and they don't offer much power. One I saw advertised for $955 featured 500 watts of power. A solar panel keeps a 12-volt marine battery charged, and your loads draw the current from this. At this price, you could buy a ton of conventional battery-powered radios, TVs, and flashlights.

The price of a generator is determined by its size and its capacity. A Yamaha EF1000, which produces a maximum of 1,000 watts with a six-hour refueling time, runs around $600. A YG6600D, at 6,600 watts, runs closer to $1,800 as of this printing. Every line of generators features a variety of models such as economy, deluxe, commercial or industrial, and low-noise.

You can run a generator and connect a heavy-duty extension cord to it. This will enable you to run some small lights, a radio, and temporarily, a small kitchen appliance. It's more convenient, however, to connect the generator to your service panel. This is done with a *transfer* that isolates individual circuits from the utility power lines. These are the circuits you want the generator to run when you lose power.

Positively Shocking

There are strict code rules regarding the use of extension cords for the temporary distribution of power. A cord cannot be run through a window or through holes in walls or be attached to a structure. Check your local codes for all applicable rules.

The transfer switch and the connection of the generator to your electrical system must be done by a qualified electrician and be done under permit!

A transfer switch plays an important role. It …

➤ Prevents the generator's power from back-feeding through the utility lines, where it can injure or electrocute utility workers repairing the lines.

➤ Prevents any electrical current from the generator from causing a short circuit with your normal house current when the power is restored. This short circuit could cause a fire or cause the generator to explode or burn.

The transfer switch isolates individual circuits with switches that allow you to replace the utility's power with that of the portable generator. You then can switch back when the power is up and running again. Each of the separate house circuits is protected by one of the generator's circuit breakers.

Bright Idea

Have your electrician show you how to use a transfer switch. You don't want to be guessing about how it works when the time comes to use it.

Read the Manual

It seems self-evident that you should read the owner's manual with every new tool and appliance you buy, but we're often lax in this department. Portable generators are a little more complex than a new toaster, so I heartily recommend reading the owner's manual completely before operating the equipment. Here are some other safety rules:

➤ Don't let children operate the generator. Keep them and pets away from it when it's running.

➤ Know how to shut off the generator in the event of an emergency.

➤ Use clean fuel and avoid contaminating the fuel tank with dirt or water.

➤ Switch the engine off and allow it to cool before adding fuel. Refuel in a well-ventilated area.

➤ Check the oil and air cleaner before each use. Replace as indicated by the owner's manual.

➤ Be sure the voltage selector switch is in the correct position (120V or 120/240V) for the current you will be running.

➤ Divide the load as evenly as possible between the generator's circuits when in 120/240V mode.

➤ Test the GFCI receptacle regularly.

➤ Turn the generator's circuit breaker off before starting the generator so the load doesn't begin drawing current until the generator is running smoothly. Turn the circuit breaker off before stopping the generator as well.

➤ Store the generator in a dry place. Refer to the owner's manual for recommendations regarding long-term storage with gasoline in the tank and changing the oil.

➤ Never operate the generator in a closed room or other area. Only operate it in open areas, in dry conditions, and on a firm, level surface.

➤ Test the generator from time to time according to the owner's manual. You want to be certain it's in working order when you need it.

Always keep in mind that switching between a generator's power and the power supplied by your utility is serious business. Follow all safety rules and explain them to your family.

The Least You Need to Know

➤ Your workshop should have at least one and possibly two dedicated circuits and plenty of light so you can work safely.

➤ A dedicated computer circuit will prevent accidental tripping (and data loss) caused by someone using the same circuit in another room.

➤ Raceways provide a convenient means of wiring around finished walls and surfaces.

➤ An emergency generator, if kept in good working order, will keep you from bumping around in the dark the next time you have a power outage.

Your Own Hi-Tech Revolution

Once upon a time, TVs were all black and white with either rabbit-ear antennas sticking out the back or a huge, roof-mounted antenna that looked like a clothes-drying rack. Phones were all rotary dial, were hard wired into the wall with four-prong plugs, and performed two functions: dialing out and receiving calls coming in. Some of the particularly tough plastic ones also were useful for anger management because they were almost impossible to break, regardless of how hard or how often you slammed down the receiver. No one had computers, modems, or fax machines, and intercom systems usually consisted of one family member hollering for another.

Now all the fantasies from movies and cartoons you viewed as a child (and probably still view) are gradually coming true. No, not the ability to get hit by an anvil dropped on your head from the top of a tall building or to be blown up with nitroglycerin and walk away intact. Rather, we have an increasing ability to automate our homes to a degree unimaginable by past generations. You'll have to decide for yourself how necessary it is for you to be able to answer your front doorbell via your cell phone while you're a few thousand miles away drinking tea in Beijing ....

Bright Idea

If you're unsure as to what extent you want to automate your home, visit a new development of homes that have been wired for automation. Talk with the real estate agent or builder about the system, and try it out. This will give you some ideas for a system of your own.

The key to limiting the amount of wiring you need to do now is not so much your selection of technology; it's building a system that can be adapted to future changes. With today's communication and computer technology, change is the one certainty (unlike, say, a computer operating system that won't crash while displaying cryptic messages on your monitor). Totally wireless systems aren't with us yet, so you'll still have some drilling and pulling to do.

The Automated Home

An automated home is defined by its electronic conveniences and safeguards. The term itself is dated and more accurately describes a home of integrated systems. The array of gadgets, monitors, and controls that can be integrated includes ...

➤ Analog phone systems.

➤ Computer networks.

➤ Fax machines.

➤ Modems.

➤ High-speed Internet access through ISDN and DSL connections.

➤ Full audio systems.

➤ Cable and satellite TVs.

➤ Security systems.

➤ Lighting controls.

➤ Sprinkler controls.

You certainly don't need to automate your house to this extent. On the other hand, many of us gradually will update our phone systems and television cable. Various business powerhouses (telephone and cable companies, Internet service providers, and hardware manufacturers) are all vying for your business. No one can predict future technology, but you can prepare for it by bringing your house up-to-date with the best available hardware and wiring.

A Wealth of Possibilities

If your goal is a thoroughly automated house, consider the following scenario:

All of your thermostats are online and can be adjusted remotely from any telephone. Sensors will call a preprogrammed phone number if the system is faulty or the temperature is dropping too low. Additional devices control your irrigation and sprinklers, entry gates, doors, and garage doors. In the event of a fire, you can shut down your furnace or air conditioning to prevent the spread of smoke. Lights can be programmed to turn on and off randomly while you're away to give your home a lived-in appearance. Automatic draperies and window coverings can be opened or closed remotely.

Your telephone system can double as an intercom so you can call specific room phones to announce that dinner is ready. You can program your system so the phones do not ring after a certain hour; all these calls then go to voice mail (with instructions on how the caller can contact you if it's an emergency). A printout of all calls will help you determine which ones to return. A door phone can serve as a doorbell, and a separate line can control the door lock. You can even remotely answer your front door (one that is wired with an intercom) using a cell phone.

Positively Shocking

It takes just as much labor to run one cable as to run two or three cables taped together. You'll only undercut yourself if you install less cable rather than more. You might not need extra wiring now, but there will be more than enough hardware in the future for it. If nothing else, sufficient cable is a positive selling point for your house.

Internet connections bring real-time audio, video, interactive games, and movies to TV screens and computer monitors on demand. If you have more than one computer, they can be connected together into a local area network (LAN).

Central control units can be programmed to run multiple systems at once. One program can run your bath water, activate your security system, turn down the lights, and turn on a separate bathroom heater before an evening bath. With the right system and programming, you could control your entire house remotely from a cell phone while thousands of miles away. (This would really endear you to your house-sitter.) While you're at home, you can use voice control as sensors pick up your vocal commands.

Features and Benefits

■ A network of transmitters and DHC switch, fixture and receptacle receivers that communicate via unique coded switching signals over a home's AC wiring to automate control of lights and appliances

■ DHC components replace switches and outlets, which simplifies installation compared with other automation systems that require dedicated proprietary wiring backbones.

■ An effective way to contribute to energy conservation and extend bulb and appliance life

■ Offers significant security benefits to today's homeowner, including programmed automatic and manual home security features

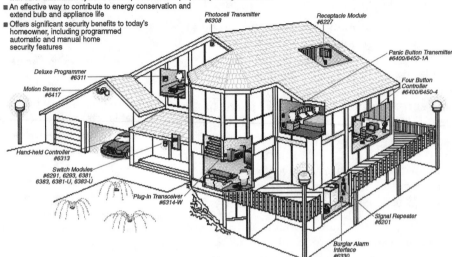

One example of automated controls from Leviton.

Electrical Elaboration

It isn't necessary to have a computer or to route your cable into a computer-controlled system to have an automated house. A programmed thermostat, for example, is an automated device that is independent of a larger, central control panel. Every system is different, and you can have several independent systems in your house if you want.

There are more possibilities, but you get the picture. You're mostly limited by your imagination and your budget. Each room can be programmed according to its use or your whims. Before you start thinking that your house will seem like a miniature version of Singapore with a dose of *Star Trek* thrown in (with monitors and security cameras everywhere), consider the advantages of automation:

➤ A camera mounted near an entry door can tell you instantly whether a friend or a solicitor is ringing the bell. You also can talk to the person through a speaker and let him or her know you'll be right down (or not)—a big plus if you're far from the door.

➤ A programmed phone system greatly expands the use and accessibility of your phones beyond immediate conversations.

➤ Small nuisance details, such as turning on the sprinklers or activating the security system, will take care of themselves.

➤ You can avoid redundant purchases (such as VCRs or printers) because you won't need them in multiple locations. One VCR, for example, can be shared by multiple television sets.

Although battery backup prevents most systems from losing their programming, one thing remains true: You still need electrical power and cable connections to the outside world to make most of these systems work.

Decisions, Decisions

Competing communication technologies have one common purpose: They all bring signals into your home and then move them around from one device to another. Once you've decided on a system(s), it's up to you to install the devices and any wire or cable necessary to connect them. You'll be choosing from the following technologies:

➤ Twisted copper wire

➤ Coaxial cable

➤ *Fiber-optic* cable (more accurately, optical fiber cable)

➤ Wireless

Copper is the old standby that's tied telephone systems together since Alexander Graham Bell first phoned for his assistant, Watson. It's even given rise to the term POTS (plain old telephone system). One of copper's limitations is the amount of signal it can carry (bandwidth), but clever engineers have come up with ways to enhance and improve its performance.

Coaxial cable was first used by cable TV companies to bring all those additional channels into your home that you never knew existed (and never watch anyway). It has evolved into a broader role of communications within the home, creating networks among television sets, VCRs, laser discs, and computers.

Ask an Electrician

Fiber optics is the branch of the science of optics that uses optical fibers to transmit light. Optical fibers can transmit light through straight lines and around curves with no noticeable change in the light's intensity.

Fiber-optic cable consists of strands of spun glass or plastic wire or fiber and an insulator. Information is sent along fiber-optic cable as light pulses at a far greater speed and in a greater capacity than old copper lines. Most long-distance phone signals are now carried along fiber-optic cable.

With the exception of wireless, these systems rely on some very low-tech skills: drilling and pulling. Even if you don't completely understand the underlying technology behind the different systems, you do understand how to run the wiring. The hardware and receptacles should simply be where you need them. At least in the case of telephones, however, there are a few NEC rules. Article 800 of the code states the basics:

➤ You can't run telephone wiring in the same holes or raceways as electrical wiring.

➤ Telephone wiring should end in a separate box and should not share a box with electrical conductors unless the conductors are separated by a partition.

➤ The wiring must be grounded. (This is done by the phone company.)

Positively Shocking

The purpose of the code is to protect you. If you were to violate the code and run electric and telephone wires in the same raceways, for example, there would be a lot of static in the telephone lines (caused by the electric lines). The phone lines would be bad for phone calls and would be useless for modem (data) calls.

Mixing and Matching

Futurists of all stripes tell us that communications will simply get faster. There will be no more waiting for online downloads or waiting to get online because we'll *always* be online. You can speed things up now and prepare for the go-go years ahead.

Completely wireless systems are on the horizon, but until they're perfected and the satellites are up (Iridium, a global satellite telephone and paging service, entered voluntary Chapter 11 Bankruptcy recently), we'll all be using some form of cable for our home automation.

Speed Demons and Slowpokes

Not all Internet and data connections are created equal. You might have the hottest PC and the fastest modem on the planet, but your online connection can slow you to a comparative crawl. The cables running into most homes were never intended for transmitting data, especially data at fast speeds. Copper wires transmit voice or electricity, and coaxial cable carries television signals.

The telephone industry couldn't very well tear up a zillion miles of copper wire and replace it overnight with something faster like fiber-optic cable. The industry has, however, come up with a few strategies to improve the data-transmitting capability of copper wire such as ...

➤ An Integrated Services Digital Network, or ISDN (only twice as fast as current modems).

➤ T1 lines, which initially were developed for multiple voice connections for commercial customers (very expensive, very fast).

➤ Digital Subscriber Lines, or DSLs. Through the use of computers at each end of the copper phone wire, these lines achieve higher rates of data transfer (the middle choice in speed and cost).

Bright Idea

Even if you don't want the expense of a DSL, consider a second phone line anyway. This gives you clear Internet and fax access without sharing a single line for voice use.

DSLs are hot right now, and in some areas, phone companies are scrambling to keep up with residential demand. More affordable than a T1 line, the copper wire from a house typically is run to a central office of the phone company where it connects to a switch. An asymmetric digital subscriber line (ADSL) is a common residential version of this technology. Asymmetric refers to the two-way nature of the bandwidth. This allows for most of the bandwidth to be devoted in a downstream direction toward you, the user. Why? Because that's the direction most Web data flow, especially graphics or software downloads, which would take forever if they didn't have extra bandwidth devoted to them. Most of what you send upstream consists of text messages such as bad jokes you e-mail to your friends and the occasional Web browser command.

Electrical Elaboration

The terms **twisted wires** and **twisted pairs** take on some significance beyond the fact that they are physical descriptions of wires. Wires carrying a current produce an electromagnetic field. These fields can interfere with voice and data transmissions, so it's best to reduce them as much as possible. When two wires are running parallel to each other, their fields are disruptive. If the wires are twisted around each other, the fields tend to cancel each other out. The higher the numerical rating, the more twists per inch the wire has and the greater its capability to reject interference. Those twists are an important part of the cable's design!

Phone lines (as they traditionally have been used) are slow for data transfer. A telephone converts your voice into an electronic analog signal. An analog signal doesn't make the best use of copper wire for data transfer because it has to convert digital information to analog form and then back again via a modem. A DSL transmits digital data directly without going through any analog changes. This allows the phone company to squeeze more data into the copper wire.

In fact, you have the option to use the bandwidth for both analog and digital transmission at the same time on the same single DSL. This usually requires that a signal splitter be installed at your home, which means an expensive visit by the phone company technician. A new standard, referred to as *G.Lite* or *DSL Lite,* is a *splitterless DSL* (because the splitting is done at the central office). This is a slower version of ADSL, but it can be installed less expensively.

Coax Dance

Not to be left out in the data-access cold, cable companies have jumped into the fray by enhancing their coaxial cable infrastructure with fiber-optic lines. This has enabled cable companies to carry TV, data, and telephone signals over their lines.

Cable companies faced the same issues as telephone companies: They had a lot of existing cable in the ground, and tearing it all out and replacing it with fiber-optic cable would have been costly. As a compromise, they have run fiber-optic cable to clusters of homes, at which point it connects up with existing coax. This fiber-to-the-curb approach has worked for now, but some new homes are being wired for complete fiber-optic cable by owners with an eye toward the future.

Electrical Elaboration

Fiber-optic lines carry data in megabits per second. A beam of laser light shines down the fiber, following all the twists and turns until it's detected at the other end. The pulses of light from the laser (which is turned on and off billions of times per second) are converted into an electrical signal at the receiving end. (The data is in binary numbers that are written using only zeroes and ones, so "On" could mean one and "Off" could mean zero.)

A cable connection is always "On." You don't have to log on to the Internet and wait for a telephone connection. It also offers faster transmission than high-speed phone service, and a cable modem can connect to more than one PC in your house. Eventually, we'll be running cable to an array of new "smart" appliances such as refrigerators that will monitor their own contents and place an order with the grocery store (well, only if you want your kitchen to get this weird).

DSL advocates see at least one problem with cable modems: Your entire neighborhood shares the coaxial cable, so the bandwidth must be shared. As the number of users increases, the cable operator must run more cable to maintain enough bandwidth.

Look Ma, No Wires

We're all familiar with orbiting satellites, one form of wireless communication. There also are ground-based wireless systems such as those used by cell phones. They use radio signals from base stations. Each base station covers a small area and links to the next base station. Today, cable still appears to be king. Given the choice, wireless is more convenient because it eliminates a lot of wiring, but it's not yet affordable.

All areas of technology are changing rapidly. The demands of today's Internet and phone users will increase tomorrow. A system installed for today's user might have a short life, but the infrastructure cost could be huge. Given the expense and uncertainties, you might not end up with the best system because your options depend on what companies are willing to provide in your area. The bottom line usually is how much it costs them, not what it can provide you.

Hedging Your Wiring Bets

Instead of agonizing over which technology to choose, keep it simple and choose them all. It requires little additional trouble or overall expense to run wiring for phones, TV, sound, Internet, and networking. In new-house construction, it's obvious that you should run as much wiring as possible while the walls are open. Remember, one reason to run all this cable is to connect equipment from one part of your house to equipment in another part so you don't have to duplicate equipment. A VCR in your living room can be connected to televisions in all the bedrooms, saving you the cost of extra VCRs. Intelligent wiring will keep other expenses down as well. (Of course, if you feel you need more than one VCR, that's your choice.)

It isn't necessary to install a separate, specialized wire or cable for each phone or computer hookup. One cable, *Category 5 cable*, is multipurpose. It's not the only cable you should consider installing, but it will keep you current for the time being.

Staying Alive with Cat 5

Category 5 cable (Cat 5, for short) consists of one to four pairs of twisted copper phone wires, and it is used to connect LAN equipment (computers and printers, for example). It also can carry telephone voice and data. It has two advantages over standard phone lines:

➤ It can carry data at very high speeds.

➤ Cat 5 offers more bandwidth (more data at once).

Relatively few home computers currently operate fast enough to take full advantage of Cat 5's speed or bandwidth, so it will be awhile before it becomes as outdated as standard telephone wire. In addition to computer networks, Cat 5 also is suitable for video. This is very much the cable *du jour* for the automated house. At a cost of roughly 10¢ a foot, you should install two Cat 5 cables in each run. Just repeat this mantra: You can't have enough cable in an automated house.

Module Selection Guide

JACK	CABLE	CAT. NO.	APPLICATION
Category 5 Data Grade Jack	Category 5 UTP Cable	40838	Ideal for all types of data and voice wiring including phones, computer modems, fax machines, local area networks (LAN) and home automation.
6-Conductor Voice Grade Jack	Category 3 UTP Cable	40836	Phones, computer modems and fax lines.
F-Connector Jack	Coaxial	40831/40731	Used for TVs, cable televisions, VCRs and satellite systems.
Banana Jack	Speaker Wire	40837	Most often used for home and office audio/video systems.
Binding Post Jack	Speaker Wire	40833	Home and office audio/video systems.
RCA Jack	Shielded Audio/Video Cable	40830	Home and office audio/video systems including VCR, cable TV, satellite systems and connection to other TVs.
BNC Jack	Shielded Video & Data Cable	40832	Commonly used for closed circuit TVs, high end video systems & local area networking systems.

Leviton low-voltage jacks.

Cooperating with Coaxial Cable

Coaxial cable is the standard for cable TV and video transmissions. You should select the best grade of cable available when wiring your home (for example, Belden #9114). This means quad-shielded RG 6 coax, which has a heavy insulation that gives you a clearer picture. Prices vary depending on the grade of RG 6, but you can figure 20 to 25 cents a foot, depending on the supplier. Internet prices might be less.

Fiber-Optic Cable

Some builders are jumping ahead and wiring with fiber-optic cable for even greater speed and bandwidth than Cat 5. It's a more expensive cable, and it also takes more expertise to install. Do you need it? Probably not, but it most likely will be the standard at some point in the future. You can prepare for this eventuality when you install your Cat 5 and coaxial cables by pulling fiber-optic cable along as well.

Package Deal

If the prospect of installing your own home-management wiring and hooking it up is a little daunting, you can purchase a ready-to-go system from Lucent Technologies (www.lucent.com). The HomeStar Wiring System, designed by Bell Labs, has a single infrastructure that encompasses everything you can do with Cat 5 and coaxial cables. The system consists of one or more distribution centers or control boxes (the Service Centers) and a series of outlets. Each center can support 16 outlets, and the entire system can support up to four centers or a maximum of 64 outlets—more than enough for most residential use.

Positively Shocking

Just because you or a builder installs Category 5 cable doesn't mean you have Category 5–qualified circuits. The cable, connectors, connections, and testing of the circuit must be certified Cat 5 as well. Be sure the supplier's literature states that these items are certified Cat 5.

Ask an Electrician

Coaxial cable, invented in 1929, consists of an inner conducting core surrounded by insulation and a second insulated conductor that acts as a ground. It can carry multichannel telephone signals and TV transmissions, and it can be used for computer networking.

What does all this cost? Lucent estimated (in 1998) that an installed HomeStar System would run approximately $700 to $2,000. It's one-stop shopping, but it's not necessarily a bargain compared to buying separate components and cable. With off-the-shelf systems, you can get by with one distribution panel if you have more than 16 outlets.

Positively Shocking

You can buy a combination cable that includes coaxial and Cat 5 cable, but it will cost more than the combined cost of the individual cables. One Internet supplier quoted $1.39 a foot for its brand. It's simpler and cheaper to tape the ends of two Cat 5 cables and one coax cable together and proceed with installation.

Bright Idea

Before going crazy and installing a wireless system all over your huge, rambling house, install one or two light controls and see how well they work. Buy a minimum number of controls and use them at different times of the day (for example, when a lot of loads are operating and again when few are running). If you're satisfied, continue installing controls.

Not to be left out of the perceived-to-be-explosive home-wiring market, FutureSmart Networks (www.futuresmart.com) offers its Residential FutureProof Interactive Network, and IBM (www.ibm.com) steps up with the authoritative-sounding Home Director. Homebuilders, with a not-so-subtle push from hardware manufacturers, are getting into the picture and prewiring new houses.

X-10, an Original

X-10 is the original automated house protocol, designed to communicate commands (On, Off, Dim) between X-10 transmitters and X-10 receivers. It requires no special wiring and sends its commands along standard 110-volt electrical cable. When a command is sent to a specific receiver, the command carries a unit identification number so it will be ignored by other receivers. The system allows up to 256 different addresses to be programmed into the transmitter. It only controls electrical devices, not phones or data.

X-10 (a trademark of X-10 USA) has been in use since 1970, and it is installed in more than four million homes worldwide. It's a basic system that controls lighting and appliances, and it often is installed by do-it-yourselfers. In the opinion of some observers, it is not as reliable as systems based on dedicated Cat 5 and coaxial cables because a system that uses power lines is subject to interference from electrical current.

Other (presumably improved) systems have followed X-10, including CEBus and LonWorks. CEBus requires no central control panel because the network intelligence is contained in the individual CEBus appliances. These appliances can use electrical cable, coaxial, fiber optics, infrared signaling, or any variety of twisted-pair wires to communicate with each other.

The Installation

Think of automation as home-management wiring as opposed to electrical wiring, although it follows some similar principles. Instead of providing power to various loads, you're going to be controlling these loads remotely or through preprogrammed commands. You won't lack for suppliers of both cable and hardware who want to help you automate your home.

Just as an electrical system has a service panel, your phone, TV, and connectivity wiring will have one main distribution panel (and possibly one or more smaller panels depending on the size of your house). One distribution panel accommodates cable for voice, data, LAN, and television transmission. Once you've decided where to put the panel, you'll run each line out of it just like your electrical branch circuits.

Staying in the Closet

Unlike with electrical wiring, the NEC doesn't require you to locate your distribution panel for connectivity, phone, and TV cable as close as possible to their entry points into your house. In fact, many people install their panels in an accessible closet or cabinet. All of your cable runs should be connected to your panel in a "star" wiring pattern (one cable per device). A star pattern makes it easy to isolate problems when they occur.

Distribution panels vary by size, price, and how many connections they can distribute via hubs. Hubs act as an interface between the incoming cables (coaxial, phone) and the cables running to the various outlets and devices.

Snap-in modules support phone, data, LAN, audio, video, lighting, HVAC, and home-security systems. These connections support specified numbers of telephones, computers, VCRs, and television sets. Prices run from less than $200 to more than $500. (Buy one with room for future expansion.) Note: A distribution center for Ethernet connections requires a 120V house branch circuit (if you're only connecting two computers, you can just use an Ethernet crossover cable).

Cable is much less expensive than your labor, so leave an extra six to eight inches of it in the box for each run.

How much room will a panel take up? One manufacturer lists three panels at 14, 28, and 32 inches high. All are 14.25 inches wide and 4 inches deep. The larger ones have more capacity, of course. A panel that sits flush between the wall studs will give you a neat, clean installation.

Positively Shocking

Be sure to label each cable going into the wire closet; otherwise, you'll never be able to sort them all out. Masking tape leaves a mess on the cable insulation, so try an address label folded around the cable instead.

Jacks Here, Jacks Everywhere

Our homes might well be full of Internet appliances in the years to come—appliances that are smaller than computers yet have full Internet and networking features. If they're cheap enough, we'll buy them by the dozen and spread them all over, one or two in each room. Unless you've gone wireless by then, you'll need a jack or a receptacle to run them.

That said, your safest bet is to run cable into every room, even to opposite sides, so you have two or more viable locations in each room. This requires one standard single-gang plastic electrical box per location. A blank cover plate can be installed on any unused boxes. Use the same techniques to install these cables, as you used with electrical cable, with a few precautions:

➤ All cables should be in the star pattern, connecting to the distribution panel.

➤ You must use the appropriate connectors for the specific cable you're installing.

➤ Boxes should be installed at the same height as other electrical boxes but not on the same studs.

➤ Coaxial, telephone, and Cat 5 cables should be installed away from existing electrical wires. (Twelve inches is recommended.) When they must cross electrical cable, do so at 90-degree angles.

➤ Handle these cables more gingerly than electrical cable. Their insulation isn't as tough, and you can damage the cable if you force or yank it through the walls or floors.

Bright Idea

Many technicians recommend installing plastic conduit from the distribution center to the various boxes in addition to your dedicated wiring. This leaves the system open for future changes without having to run new cable and tear into the walls.

Unless you're pretty technically oriented, you'll need to call or at least consult an electrician when considering dedicated wiring. You can save some money by planning out the locations of boxes and the distribution center and then pulling the cable yourself. Entire books have been written on this subject (although some simply discuss what the wiring can do, not how to install it), and even electricians sometimes call in specialists, especially for complicated home-theater setups.

Is It Worth the Trouble?

Some hardware manufacturers have convinced builders to install, or at least offer, dedicated wiring systems for home automation. They suggest that more and more buyers (and certainly future buyers) will demand systems for high-speed Internet access, multiple telephones and televisions, and computer networking. It isn't as though you won't be able to sell your home if all you have is an old copper phone line and electrical wiring, but it does suggest the direction that home wiring is going. Even if you're a technological Luddite who doesn't own a TV, it's a good idea to install this additional wiring while you're upgrading your electrical system. Drilling another set of holes isn't that big of a task, and you'll be glad you did it if you decide to become a technophile.

If your budget is limited, you can just run the wire and install the boxes and leave the distribution center for another year's budget. Just be sure all the wires are clearly marked for later connections and are protected from damage. Depending on the number of wires, I would run them inside a three-gang plastic box and then screw on a blank cover plate.

The Least You Need to Know

➤ Just as electrification came into homes a hundred years ago, automated "smart" homes are part of the next wave of technological change.

➤ Both telephone and cable companies offer versions of faster Internet access; weigh your choices carefully before signing up for one or the other.

➤ You can't have enough dedicated cable in your house, so install plenty in every room.

➤ Some manufacturers offer complete dedicated wiring packages. Always compare their costs against off-the-shelf products.

➤ Installing a raceway or conduit will enable you to change your system and to adapt to future technological advances.

Alarms, Detectors, and Security

Anne Morrow Lindbergh once said, "Only in growth, reform, and change, paradoxically enough, is true security to be found." That's probably true, but a good door lock can help people for whom change is a little slow in coming. Any number of manufacturers and suppliers have a security system for you. Systems range from fake window decals that probably wouldn't fool a four-year-old to elaborate, monitored systems that could detect a dust bunny at 20 feet.

A security system need not be at the top of your safety list, but smoke detectors should be. There's no question that this relatively inexpensive investment is a lifesaver, and this is one of the reasons the installation of smoke detectors is mandated in new construction. You have several options to choose from when installing smoke and heat detectors, so planning is critical.

Security systems use low-voltage wiring, which requires a standard 120V circuit and some form of transformer (to step down the current to a lower, usable voltage). In addition to the transformer, a major difference between low-voltage and regular-house-voltage wiring is the size of the wire. Low-voltage systems often use 18 AWG wire (also known as bell wire or thermostat wire). It's very thin and can be damaged easily if not handled carefully.

You should plan your low-voltage circuits just as carefully and methodically as your other electrical circuits. You want to use the most efficient routes and maintain a distance from your other electrical cables as they pass through the walls and floors. This chapter starts with a short primer on low-voltage wiring and then goes into the systems themselves.

A Class 2 Act

Most low-voltage residential wiring is the Class 2 type. It carries such a low voltage that it's not considered to be a fire hazard or a notable source of electric shock. It should not be run in the same holes or raceway/conduit as electrical cable because doing so could damage it. Class 2 wiring, however, can be attached to the outside of the conductor raceway (using straps or tape) that supplies power to the specific device or appliance that the wiring connects to or controls. A thermostat wire, for example, can be attached to any conduit or raceway that supplies power to a furnace. Class 2 wire cannot be used in the same box or enclosure as electrical cable unless there is a physical barrier that separates the two types of cable.

Eighteen-gauge, Class 2 wire is relatively inexpensive (it's around a dime a foot depending on the number of conductors), so think about running it into any room where you might consider installing an intercom or a security sensor. As long as you're opening up walls and drilling holes, follow the philosophy that you can't have enough wiring. Even if you don't use it, a future owner might, and you'll make it that much easier to sell your home.

Alarm Systems

Despite FBI statistics that indicate a decrease in major crimes the last few years, there is no lack of residential security systems or their promoters. Systems range from fake window stickers (that supposedly fool illiterate burglars into believing you have a real alarm system) to high-end, multiple-sensor, hard-wired arrangements that any Federal Reserve Bank would be proud to have. Your system can be as simple or as elaborate as you want, but keep in mind the following caveats:

➤ Good security is based on diligence. This means locking doors and windows regularly, watching your neighborhood, and providing sufficient outdoor lighting.

➤ Even the best system is worthless if it isn't used regularly.

➤ A cheap system will give you exactly what you pay for—not much.

➤ A monitored system is more desirable than an unmonitored system.

A security system, along with smoke and heat detectors, is primarily a means of protecting human life. You can always replace most possessions. You want a system you can depend on whether you install it yourself or use a professional. Each supplier is biased in favor of its brand of alarm, and it's doubtful you'll hear the same advice twice.

To Monitor or Not To Monitor

A monitored security system is tied to a central reporting facility through your telephone line. Operators will report your alarm to the local police or fire department as appropriate. They will call you at home or at work to confirm the alarm. With some systems, if your phone line is cut, your alarm is automatically reported through the use of another customer's line. You pay a monthly fee for this service.

An unmonitored system depends on your neighbors or a passerby to call in the alarm when your siren or horn starts blaring. This is far from dependable. It's a rule of urban life, for example, that car alarms are ignored as background noise. Do you think you'll be more fortunate with an unmonitored house security system?

Basically, an alarm system informs a would-be burglar that he (it's almost always a he) needs to get in and out really fast or try next door instead. A monitored system at least gives you some chance of recovering your goods.

Electrical Elaboration

Motion detectors are fairly sensitive, as they should be. We kept a helium-filled birthday balloon in the living room of our last house, not noticing that it was slowly descending day by day. Eventually, the motion detector picked it up, as did the system's monitoring station and the Seattle police, who were arriving at our house as I was driving in. It was embarrassing, and we learned a lesson, but we really wasted the time of two police officers.

False Alarm

Police departments all over the country have a statistic they absolutely loathe: the number of false alarms they answer every year. One report from the Chicago Police Department indicates that, out of the 300,000 burglar alarms they respond to every year, 98 percent are false alarms. Police departments have better things to do than respond to false alarms. These same police departments have passed ordinances governing the installation of alarm systems and fines for false alarms.

What is a false alarm? To the police, it's an alarm that has been set off for reasons other than criminal activity or an emergency. The main cause of a false alarm is human error including ...

➤ Using the wrong code.

➤ Misuse of the system by people not familiar with it.

➤ Failure to close and secure all windows and doors.

➤ In the case of a monitored system, forgetting your password when the security company calls after the system has been set off.

➤ Improper installation of the system or some of its components.

➤ Faulty batteries or equipment.

Positively Shocking

Some cities require that you obtain an alarm permit before installing a security system, especially a monitored system. Your monitoring company might need your permit number on file if a police response is requested. This permit is required whether the system is installed by a homeowner or by a professional. Check local regulations before doing any installation.

The police and your alarm company would prefer that you never have a false alarm, and they have all kinds of helpful advice for you:

➤ Make sure that you and your family are completely familiar with your security system.

➤ Secure all doors and windows before turning on the system.

➤ Make sure any changes in your decor or furnishings don't affect the alarm sensors. (This includes new pets, which can set off a motion detector.)

➤ Cut down on drafts that can blow a curtain into the range of a motion detector.

➤ Replace the backup batteries every three to five years.

➤ Do a monthly system test.

Hard Wired or Wireless?

Elaborate security systems have a hard wired contact at every window and door. Switches are inserted flush into the jambs (the frame into which a door or window sash closes), and corresponding magnets are installed in the doors or sashes themselves. If the system is armed and a door or window is opened, the contact is broken and the alarm is set off. This system is practical if you have open walls during new construction or remodeling, but it's much harder if your house is finished. Older homes often have a lot of windows, and double-hung windows—those with two movable sashes—would need one contact per sash. That's a lot of wiring and a lot of contacts.

Like any other electrical system, a wired system requires that cable—in this case, Class 2 wire—be run from a monitoring device to every sensor. A wired system should have a battery backup in the event of a power failure. One advantage of a wired system is that the sensors either are hidden from view or are smaller and less noticeable than some wireless systems.

Wireless Is Almost Effortless

Wireless systems run on an FM radio signal and work very well. You can choose between a professionally installed and monitored system or off-the-shelf, homeowner-installed components. There are plenty of choices to fit every budget. Every sensor requires a battery, which must be monitored. Wireless advocates suggest that a three- to five-year battery life is normal.

There are plenty of wireless systems to choose from, but the more sensors you use, the more the system will cost. Some typical features of a wireless system include ...

➤ A 12V DC power supply with a 120V AC adapter.

➤ Battery backup power.

➤ A power and arm/disarm indicator.

➤ A low-battery indicator.

➤ Indicators for gas, smoke, and fire detection.

➤ Some type of audible alarm or siren.

➤ A panic button.

Wireless systems are extremely reliable. Transmitters in the sensors are supervised. That is, your control unit always is aware of the status of each sensor, including its battery power. If you have a monitored system, the control unit will contact the monitoring center to alert them as well. Control units also will indicate whether a door is open or closed.

Many wireless systems come with remote, wireless keypads, some of which are so small that they double as a key ring with simple "On/Off" controls. Larger keypads include numbers for entering an alarm code and arm, disarm, and emergency commands. Smaller versions of the keypad are available for secondary exits or bedrooms. Some manufacturers offer central monitoring for do-it-yourself-installed systems.

Electrical Elaboration

Different security systems use different means of detecting noise and movement. A detector picks up the sound of broken glass with a sensitive microphone. An infrared detector picks up a change in temperature as a body enters an area. A photoelectric detector uses a light transmitter that sends an infrared beam to a receiver; if the beam is broken, the receiver sounds an alarm.

X-10 System

X-10, as you might recall from Chapter 24, "Your Own Hi-Tech Revolution," is a system that sends commands between X-10 transmitters and X-10 receivers along your 120-volt electrical cable. An X-10 security system does the same thing using door and window sensors, motion detectors, and a central control unit to pick up the signals. Some systems will dial a phone number and leave a message if the alarm goes off and will begin flashing select house lights on and off.

Positively Shocking

Be careful what you wish for in security systems—you just might get it. Consider just how many monitors, bells, buzzers, and beeps you want to put up with. If it's your first security system, you'll find it takes time to get used to setting codes and shutting the system down when you enter, but eventually it becomes second nature.

Driveway Alarms

For people who want an early warning when relatives show up, a driveway alarm is just the thing. A transmitter is installed several feet above the ground within a manufacturer-recommended distance from the driveway. A battery-operated infrared sensor detects a vehicle or another object coming up the driveway and sends a high-frequency radio signal to a receiver inside your home. Unfortunately, these sensors cannot distinguish one type of vehicle from another, but a surveillance camera might just do the trick.

We're Watching You

In the last few years, some parents began installing surveillance cameras to keep an eye on everybody in the house. In at least one case, a mother (and a dad, too, much to his regret) got more than she expected when the camera picked up the nanny and the dad playing house at lunchtime. Some of these wireless cameras are disguised as …

➤ Clock radios

➤ Boom boxes

➤ Smoke detectors

➤ VCRs

➤ Framed artwork

➤ Coffeemakers

➤ Stuffed animals

I bet this will make you think twice the next time you pick up a teddy bear at a friend's house. Nevertheless, a surveillance camera might have a place in your security arsenal, perhaps at the front door. The systems vary. Some wireless cameras transmit to a compact receiver connected to your VCR for ongoing recording. Others run off a branch circuit and are wired to a separate viewing monitor. More elaborate systems use motion-activated cameras that can be viewed or recorded on a remote computer.

If you really want to be cheap about it and at the same time really annoy visitors, you can install a dummy camera that does nothing but look forbidding.

Some Oddball Alarms

There's a consumer product line for everyone, and security systems prove this maxim to be true. One system, according to its manufacturer, uses microwave radar technology and can detect motion through walls, doors, and windows up to four inches thick. It also can detect these motions within certain distances. If an intruder comes within 30 feet of the sensor, which is plugged into a standard 120V receptacle, the sensor begins barking like a German shepherd. This alarm will really endear you to your mail carrier.

For a single entry door, such as in an apartment or a condominium, you can install an independent, programmable motion detector that will sweep the entry area. This type of device comes with a security code and a short entry and exit delay before going off. A clever burglar, of course, would simply rip the unit off the wall and toss it out the window unless the batteries could be removed quickly. Still, it's not a bad idea if it can be mounted out of reach.

Electrical Elaboration

You want a system that will be taken seriously. In other words, you don't want a false sense of security from an inadequate system. Amateur-looking equipment and installations simply tell intruders that they have little to worry about. False alarms from an unmonitored system simply tell your neighbors not to bother calling the police. Install a quality system and then test it as an intruder would. You don't have to break any glass or damage your house, but open unlocked doors and windows from the outside and test the response time before you rely on the system.

Another advertised alarm detects small changes in air pressure that suggest a window or a door has been opened, thus setting off a siren. Individual door alarms, used when you're in the house, will shriek when the door is opened. (These aren't a bad idea in motel rooms when you're traveling, by the way.) Finally, on our list of inexpensive alarms, a sensor that attaches to a pane of glass will sense the vibration when the glass is broken and start shrieking. These are most appropriate for windows and doors with single sheets of glass, which are susceptible to break-ins.

All of this independent hardware sells for less than $100, and some (such as the door alarms) are advertised for less than $10. They typically are available at hardware and home-improvement stores. They can be effective if someone is around to hear them and if the batteries are in working order.

Installation Issues

Both wired and wireless security systems need to be well-thought-out. You have to decide …

➤ Where to locate the central control unit.

➤ How many sensors to install.

➤ Whether to use magnetic sensors on every door and window or to use motion detectors to cover entire rooms.

➤ Whether you want a monitored or unmonitored system.

➤ The scope of work the system should do (security, fire and smoke detection, carbon-monoxide detection, and so on).

The more sensors and detectors you use, the more wiring you have to run or batteries you have to maintain. A hard-wired system requires Category 2, 18 AWG conductors that should not be run in the same raceways, conduit, or holes as electrical cable. The central control unit should be in a convenient area.

Remember, if you have pets that have the run of the house during the day, you have to adjust the system around them. A cat jumping up on a cabinet or onto the back of a couch can set off a motion detector. A zoned system enables you to adjust different areas of your house to special circumstances such as pets. Depending on the cost, you can install a system with a single zone or with dozens of zones.

Window sensors are tricky, especially in older homes with a lot of windows. If any of the various pop-culture laws of inevitable consequences are true, the one window on which you don't install a sensor will be the one someone uses to gain entry to your home. On the other hand, how badly do you need to arm your second- and third-floor windows? Better solutions for inaccessible windows are motion detectors to monitor entire rooms or glass-breakage detectors.

Monitoring costs can vary. They're usually around $20 a month or less. Local providers argue that they will respond faster and more accurately than out-of-state monitoring centers, but only statistics can tell, so check out the statistics on providers in your area. We had two neighbors near our last house with unmonitored systems. Both went off on different occasions, and no one called the police.

Look at it from a burglar's standpoint: He has a set amount of time to gain entrance, run into a bedroom, and make off with jewelry, cash, and so on. That set amount of time is the 30 to 60 seconds (times vary) that your system gives you to punch in a code and disarm it. A monitored system at least gives you some chance of recovering your stolen items. An unmonitored system probably works best if you live in an active neighborhood. In the end, a professional thief will get what he wants, system or no system. Basically, a security system just keeps the amateurs at bay.

Bright Idea

Some central control units are plugged into a branch circuit. Because the units usually are installed in a utility room or closet, the installer must run the plug and cord along the surface of your wall and baseboard, and this can be unsightly. Consider installing a receptacle right next to the control unit for a neater installation. Your installer will thank you, too.

Smoke Detectors

Every fire department in the country has statistics about residential fires and the importance of smoke and heat detectors. Smoke inhalation will get you before a fire will, especially if you're sleeping. You'd be hard-pressed to find new housing that doesn't have smoke detectors mandated by local codes.

Smoke detectors either are hard-wired to your electrical system or are stand-alone, battery-operated models. The best hard-wired detectors have a battery backup. There are two flavors of smoke detectors based on how they detect smoke: photoelectric detectors and ionization detectors. Each has its advocates, and you should consider the benefits of both types for your home.

Positively Shocking

Check with your local building department before you install your own hard-wired smoke and heat detectors. It might be a requirement for these detectors to be installed by a qualified contractor or technician.

Wireless or Hard Wired?

Wireless detectors are stand-alone units powered by 9V batteries. All UL-listed, battery-operated detectors must have a sound-emitting signal to indicate when a replacement battery is needed. These units are easier to install than hard-wired detectors, but they also require regular battery replacement. Too many people ignore the battery warnings and end up with ineffective detectors.

Hard-wired detectors are connected to a branch circuit. The best system has the detectors wired together in tandem like a series of receptacles. (Better yet are detectors connected to a monitored alarm system.) When one detector is set off by smoke, all the connected detectors also will go off, an important consideration in two- and three-story homes. To reduce the system's dependence on electrical power, detectors with battery backup are best. Detectors wired into an alarm or security system automatically have a battery backup. Hard-wired detectors can be directly wired to an electrical box or branch circuit, or they can be plugged into a receptacle.

Electrical Elaboration

The National Smoke Detector Project, a 1994–95 government study, calculated that 16 million detectors were defective. The main defects were dead batteries; vents clogged with dust, dirt, or grease; and sensing chambers filled with insects. These sensing chambers could not always sense smoke. Regular battery checks and vacuuming of detectors are recommended. Safeguard Industries recommends using its UL-listed Smoke Detector Tester, an aerosol product that the government used during its survey of smoke detectors.

Photoelectric or Ionization?

As with everything else in the electrical world, you have a choice of smoke detector technologies. Each has its advantages and drawbacks, and as usual, you get what you pay for. Some authorities recommend installing both types in your house.

A photoelectric detector emits a beam of light through a detection chamber. When smoke enters the chamber, the light scatters and is reflected onto a photo cell that triggers the alarm. Photoelectric detectors are better at sensing smoldering, smoky fires than fires with low amounts of smoke. For this reason, it's best if this type of detector is equipped with a heat sensor. These detectors most often are hard wired and generally last longer, and cost more, than ionization models.

Ionization detectors contain a tiny amount of radioactive Americium-241, which emits ionized alpha particles inside an ionization chamber. The particles allow a current to pass between two metal plates. The ionized particles are attracted to any smoke that enters the chamber. The detector senses this change in electrical conductivity and sounds its alarm. Ionization detectors are very sensitive and respond more quickly to fast-flaming, low-smoke fires. Most are battery-powered, stand-alone units, which means one might go off but others have to wait until they sense the smoke before sounding their respective alarms.

The National Fire Protection Association (NFPA) recommends that both types of detectors be installed for maximum protection.

Electrical Elaboration

Manufacturers and scientists always refer to the amount of Americium–241 in a smoke detector as "very small" or "minute." Although this is true, its presence still makes some people uncomfortable. Its 458–year half–life can seem foreboding as well. Metal plates keep the Americium at bay, and you would have to destroy the detector to get to the radioactive material. Ionization detectors should be returned to the manufacturer or to the point of purchase for disposal. They should not be thrown in the trash.

The Law

The NEC tells you how to install wired smoke detectors but not where to install them. The National Fire Protection Association Standard No. 72 has this information and is the standard used by most local codes around the country. Generally speaking, the more areas of your house covered by smoke and heat detectors the better, but at a minimum, you should install a detector …

➤ At each finished level of the house.

➤ In each bedroom.

➤ In or near all other finished rooms.

Don't install detectors in bathrooms or outside a bathroom door. (Steam can set them off and can rust the metal plates.) Extreme temperature ranges (such as in attics or un-heated garages) can set off smoke detectors unless they are temperature-rated to operate below 32°F or above 120°F. Extremely dusty areas (again, such as attics) or areas with spiders and spider webs also can set off false alarms. Smoke detectors and kitchens aren't a good match, either, because a burnt roast can set one off. (Consider a heat de-tector instead.) Manufacturers offer guidelines for installation in unfinished spaces.

Smoke and heat detectors should be located …

➤ Away from the dead air space where a wall intersects with a ceiling. (The edge of the detector must be at least four inches from the wall.)

➤ At the top of an open stairway.

➤ Near a stairway leading from a basement to the first floor.

➤ Not closer than four inches to or more than 12 inches from a ceiling when using a wall-mounted detector.

Hard-wired detectors should not be wired to a GFCI-protected circuit or to a switch-controlled circuit.

What About Carbon Monoxide?

One of the luxuries of advanced civilization is that we can elect to worry about any-thing, regardless of the degree of risk. The Consumer Products Safety Commission (Document #464) has stated that nearly 300 people in the United States die each year from carbon-monoxide poisoning (even though the population of the United States exceeds 272 million people). Your chances of dying from carbon monoxide in your home are about as good as your dog memorizing Shakespeare's sonnets. Nevertheless, you might consider installing carbon monoxide detectors.

Carbon monoxide (CO), the product of incomplete fossil-fuel combustion, is an odorless, colorless, and tasteless gas that also is quite poisonous. In your home, possible sources of carbon monoxide include …

➤ Automobile exhaust.

➤ Poorly vented gas appliances.

➤ Furnaces in need of maintenance.

➤ Kerosene heaters.

➤ Charcoal grills (especially when burning in enclosed areas).

Symptoms of carbon-monoxide poisoning include …

➤ Dizziness.

➤ Fatigue.

➤ Headaches.

➤ Nausea and irregular breathing.

These symptoms might sound like the flu, but carbon monoxide is insidious, and you'll feel far more sickly and listless as the exposure increases. The following are guidelines for selecting and installing carbon monoxide detectors:

➤ First, a detector is no substitute for regular maintenance of furnaces, chimneys, and gas appliances. Homeowners should examine chimneys for loose masonry and excessive soot and should inspect furnaces for cracks or loose panels.

➤ Only install detectors that have passed UL standard 2034. These detectors will sound an alarm when CO levels are hazardous.

➤ Install at least one detector in a common area outside bedrooms. Additional detectors can be installed near other living areas.

➤ Do not install detectors in garages, furnace rooms, or kitchens, or near bathrooms where they can be exposed to aerosol hair spray.

➤ Install, test, and maintain the detectors according to the manufacturer's instructions.

Bright Idea

Consumer Reports is a good source for both smoke and carbon-monoxide detector recommendations. Snoop around its Web site (www.consumerreports.com) for its findings.

Positively Shocking

A gas furnace might have a cracked heat exchanger, but you won't know it without an internal inspection. Not only will your furnace be inefficient, it can produce dangerous carbon monoxide. Consider a yearly checkup by a qualified technician. This also will help avoid breakdowns during cold weather.

Like smoke detectors, carbon-monoxide detectors are available in battery-powered, hard-wired, plug-in, and battery/AC models.

Smoke and Heat: Install Both?

Everyone has smoke detectors these days, but heat detectors are another matter. A heat detector senses a change in temperature, which is useful in rooms where smoke detectors are not recommended such as kitchens or attics. Some react to a specific temperature; others react to a rapid increase in temperature. There also are units that act as both heat and smoke detectors.

Heat detectors should be installed following the same precautions as smoke detectors, but you should note that they are used for different purposes and therefore should be installed in different areas of your home.

What's Safe Enough?

At this point, you might be tempted to move to a deserted island and live in a fire-proof cave (one with cable TV, perhaps). How many detectors and alarms do you need? Other than smoke detectors, I'm not sure there are any universally applicable standards. You could live in the worst part of town, leave your doors unlocked, and live crime-free your entire life. You also (like one Seattlite) could have a chain-link fence surrounding your house and two German shepherds in the yard and still get robbed. (To add insult to injury, the robber(s) loaded the goods inside the owner's new van before cheerfully driving off with the loot.)

Bright Idea

In addition to testing your security system and smoke detectors, prepare an evacuation plan for your family and pets. Do a fire drill about twice a year from all different rooms of the house to make sure everyone knows the best route out. Also make sure you know how to safely check that no one is left inside.

You must have smoke detectors. (I insist.) Whether they're hard-wire or battery, just *install them* if you don't have them now. A heat detector in a large house should be considered. A security system is up to you. Look at robbery patterns, population figures, and local crime statistics. Carbon monoxide detectors are a particularly good idea if you have an ancient furnace or a poorly drafting fireplace, or if you insist on barbecuing in your bedroom.

The Least You Need to Know

➤ Evaluate your needs before you incur the expense of a security system.

➤ As long as your walls are open or you're running other wire, install Category 2 low-voltage wire for possible alarm sensors (or intercom systems).

➤ Wired smoke and heat detectors with battery backups are more reliable than wireless, stand-alone detectors because there is less chance of human error (such as forgetting to replace the batteries).

➤ The placement of your monitors and detectors is at least as important as wiring them properly.

➤ A monitored security system will be more reliable than an unmonitored system.

More Low-Voltage Wiring

Low-voltage wiring isn't limited to telephones and alarm systems. Other equipment (such as door chimes, alarm systems, thermostats, and intercoms) also is low voltage. All such equipment requires a standard 120-volt circuit, which then can be stepped down with a transformer to a lower, usable voltage. (If you installed a door chime that actually used 120 volts, you'd think you were in the bell tower of the Cathedral of Notre Dame every time someone rang the bell.)

You already know that low-voltage wiring connects to a branch circuit via a transformer and that 18 AWG wire damages more easily than the 12/2 AWG used in many circuits. In planning your low-voltage circuits, you want the devices—the intercom stations, the buttons for door chimes—to be conveniently located. Well-thought-out low-voltage systems can bring you a different chime for every entry door, a sound security system, and a whole-house intercom. Your children can never claim they didn't hear you calling them (even though they'll probably try).

Wireless devices offer alternatives to hard-wired door chimes, intercom systems, and even thermostats. It's almost as if this book should have been titled *The Complete Idiot's Guide to Battery Power*. Hard-wired systems, however, often are more durable and don't depend on transmitters, receivers, and batteries. Power outages are an issue, but how often do they really happen? Both wireless and hard-wired devices will be discussed in this chapter.

No Escaping Mom with an Intercom

Residential intercoms have been around for many years. They can enable you to do the following:

➤ Communicate between different rooms and levels of the house

➤ Pipe music throughout your home via AM/FM radio, cassette, or CD player

➤ Monitor rooms for sound (such as a baby crying)

➤ Monitor and answer an entry door

➤ Visually monitor an entry door

Intercom systems vary from hard-wired models with master control units to those that just connect phones around your home. Wireless systems also are available.

Intercoms are a convenience and, for some people, a security appliance when used to monitor a front-entry door. You can establish two-way communication (such as at the front door, where you need it) or one-way (if you want to announce dinner but you don't need a reply from anyone in, say, the backyard).

Positively Shocking

If you have an intercom outdoors, you need one rated for outdoor use to prevent electrical shock.

Phone-Based Systems

This type of system requires a master control unit that is connected to your phone line. It can be programmed to ring all phones at once or just one extension at a time. Using a ringing sound that is distinctive from normal phone calls, you can assign a certain number of rings to individual family members who will, in theory, pick up the phone when his or her magic number comes up. I've never seen one in action, but in addition to sounding vaguely Pavolovian, I suspect it wouldn't be all that dependable with children and could be generally annoying. It also would be an additional way for your kids to torment each other by constantly ringing in each other's code.

Hard-Wired Intercoms

A hard-wired intercom wears many hats. You can use it as a public address system to be heard through all remote speakers at once or to call a single station in one room. This system also can monitor for sound and can play music. A master unit is installed (usually in the kitchen) with remote stations/speakers throughout the house. The master unit is wired to a branch circuit, and the stations are wired to the master with low-voltage wire.

Some features to look for in a hard-wired intercom system include …

➤ A built-in AM/FM radio, cassette, and/or CD player with a wake-to-music alarm function and control of the radio from intercom stations in other rooms.

➤ A surface-mounted master unit with plenty of remote stations.

➤ The capability to respond to the entry door from any intercom speaker, possibly using an electronic door *strike.*

➤ A sound-monitoring (listening-in) feature.

➤ Battery backup.

➤ The capability for intercom stations to double as panic alarms if the system is hooked up to your home security system.

Prices for hard-wired intercom systems vary depending on the manufacturer, the number of add-ons (CD players, for example), and the number of remote stations you install. Materials alone can range from a few hundred dollars to more than $1,000. Again, you can realize savings by running the wiring yourself, especially if your walls are already opened up for another electrical project. Always think ahead: If it's in your budget to install some Class 2 wire while you're installing other wiring, do so. Mark the ends of the wire so it can be identified in the future.

Ask an Electrician

A **strike** is the part of a door lock that is installed in a door jamb. Usually seen in commercial doors, electronic door strikes typically operate at 12 or 24 volts.

Wireless Intercoms

It's becoming clear that many hard-wired products eventually have a wireless counterpart. A wireless intercom requires no setup. Just remove it from its box and plug the stations into receptacles. The stations use FM radio waves to communicate. The working range of the stations usually far exceeds the dimensions of a house. These systems tend to be less expensive than a hard-wired intercom, and because they run off your house current, they require no batteries.

Video Door Phones

In addition to having an intercom speaker by your entry door, you also can install a video phone system. This enables you to see whomever is pushing the button from a remote location inside the house via a phone with a small viewing monitor. You can

Positively Shocking

If you're using an ADSL, (asymmetric digital subscriber line) for your phone communications, be sure the intercom system you purchase is compatible with this higher-speed connection. Some intercoms even allow you to talk among the phone extensions while you're logged on to the Internet. Compatibility problems are always possible.

screen your visitors visually and avoid answering the door when your local schools start sending out hundreds of 10-year-olds selling magazine subscriptions. The best systems are infrared-equipped, allowing you to view visitors regardless of the degree of ambient light available.

Ding Dong, Door Chimes

The proper name might be "door chimes," but most of us refer to them as doorbells, even if bells haven't been used for decades. Most residential devices use some form of chimes, ranging from a single note to a short tune. For people who have difficulty hearing, an adapter can be added so a light flashes whenever the button by the door is pressed.

Of course, both hard-wired and wireless versions are available. The intent of this book is to show wired systems in all their electrical glory. In the case of door chimes, they generally are more dependable because they do not rely on batteries. There also are no plug-in receivers to look at in your receptacles. Nevertheless, wireless systems have come a long way and certainly are easier to install.

Wired to Ring

A hard-wired door chime consists of the following elements:

➤ An AC power source from a branch circuit

➤ A transformer

➤ Low-voltage wiring from the transformer to the chime and then to the button

The transformer steps down the voltage. The simplest signaling system has only one push button at the front door. A second button at the back door requires a chime capable of producing more than a single sound so that each door has its own distinctive tone. In some cases, a second chime can be installed if the main-floor chime is out of comfortable hearing distance from other areas of the house.

Transformers for residential systems typically are 16 volt, are fairly small, and are wired into a junction or receptacle box with one black and one white conductor. A transformer's wattage rating generally is around 30 watts. Basements and utility rooms are typical locations for transformers.

Chimes

Our last house had older chimes that were around 24 inches long and were very loud. There was no mistaking that someone was at the door. Current models range from similar brass tubes to small xylophone-type metal plates in lightweight housing. Chimes should be mounted in a location that will enable you to hear them but not where they will be obnoxious when activated (over the breakfast table, for instance).

Positively Shocking

Transformers used for bells or buzzers use less voltage than those used for chimes. If you re-place a bell or a buzzer with chimes, you'll have to upgrade the transformer as well. Always use a transformer with overload protection (protection against short circuits).

Depending on the weight of the chimes, it's a good idea to install some support between the wall studs so you have something to act as backing for the electrical box. Some chimes are light enough and are sized so they'll screw directly into the electrical box where a device usually would be installed. The height of the unit depends on the length of the chimes. Very plain units, usually made from white plastic to blend in with the wall, can be mounted high and out of the way, especially on a tall wall. More decorative units mounted on an eight-foot wall might look better roughly at eye level or around six feet from the floor.

Wireless Chimes

Again, there also is a line of products to keep battery manufacturers happy the world over. Depending on your time and budget, these units might be well worth consid-ering, especially if you're replacing an existing doorbell in a finished house. Here's how wireless doorbells work:

➤ A battery-powered button is installed at your entry door.

➤ When the button is pressed, it transmits a signal to a remote receiver/chime that plugs into a 120V receptacle.

➤ Front- and rear-door buttons can be installed, and each will have a distinguish-able sound.

Some advertised units boast lithium batteries with an estimated 10-year life and re-ceivers with separate volume controls for around $50. Other units will even play Beethoven's *Fifth Symphony* instead of the usual ding-dong. Additional remote wireless receivers can be installed in other rooms to act as a second or third set of chimes. Basically, you're comparing hard-wired, simple technology with solid-state, electronic circuitry and batteries.

Installation is easy: Screw the button/transmitter into an appropriate location near your door(s) and plug the receiver/chime into a receptacle within the manufacturer's recommended distance from the doors. Remember to install the batteries!

Bright Idea

You can add another chime to your hard-wired system without changing the wiring. Dimango Products Corp. (1–800–346–2646) offers a battery-powered Extend-A-Bell that is advertised as being effective for up to 300 feet and can be used outdoors. A transmitter is installed on or inside an existing chime with low-voltage wire. The battery-operated chime/receiver picks up the signal when the chimes are activated.

Wiring for One or Two Buttons

For a one-button chime, remove the cover from a door chime. You'll find an interface with three terminal screws marked "Front," "Trans," and "Rear." The transformer is hard wired to a branch circuit: black (hot) to black and white (neutral) to white. In a single-button arrangement using two-wire, low-voltage cable ...

➤ A low-voltage black wire from the transformer is connected to the chime at the "Trans" terminal.

➤ A black wire from the button is connected to the "Front" terminal at the chime.

➤ The white wires from the button and the transformer are connected under a single wire nut.

No grounding conductors are involved with wiring the transformer and the button to the chimes.

For two buttons, follow the arrangement for a single-button installation, but wire the second button to the terminal screw on the chime marked "Rear." Connect the white wire from the second button under the wire nut with the neutrals from the first button and the transformer.

Troubleshooting an Existing Doorbell

Doorbells or chimes can stop working for a number of reasons. Sometimes it's as simple as a worn-out button; sometimes a wire is torn loose when other work is being done on the house. You also might have a short or another break in the wiring.

When investigating the cause of a problem doorbell or chime ...

➤ Make sure the current to the transformer is hot. (Be sure the breaker hasn't tripped.)

➤ Check the button first for loose wire connections, corrosion from exposure to weather, or metal fatigue in the spring (replace the button if the metal is too worn).

➤ Visually inspect any exposed wiring between the button, transformer, and chime.

➤ If the chime continues to ring even when the button is not being pressed, you might have a short.

➤ Using an inexpensive volt meter that will read up to 24 volts (try Radio Shack or another such supplier), test the chimes' various wire connections.

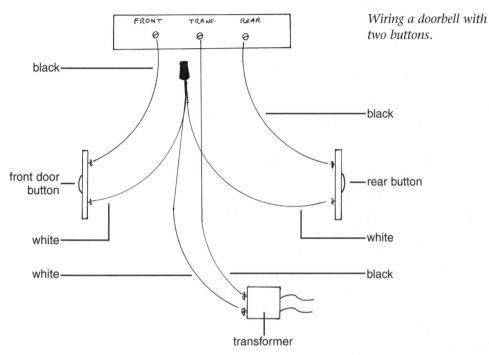

Wiring a doorbell with two buttons.

Thermostats

Modern thermostats are a wonder. Once upon a time, in the dark, old days of heating and cooling control, thermostats had to be adjusted manually. This meant you would wake up to a cold house in the winter and wonder why you weren't living somewhere fun and warm like southern California. These days, we have programmable thermostats that are relatively inexpensive and are far superior to nonprogrammable models.

Remember that a thermostat is a type of low-voltage switch. A heat sensor within the thermostat expands and contracts as the temperature changes, telling the heating/cooling system to start up or shut off. The electrical current supplying the thermostat first goes through a step-down transformer to reduce the 120 volts to something lower (often around 24 volts).

Installing a replacement thermostat is relatively easy, as you'll see in the next section.

Positively Shocking

Make sure that any replacement thermostat you buy and install is compatible with and is the correct voltage for your system. They must be the same; otherwise, the new thermostat won't work properly if at all. If you have to replace the transformer for any reason, make sure its amperage rating matches the thermostat as well.

Thermostat Replacement

Your first step, as always, is to cut off the power to the device on which you'll be working. Find the circuit breaker or fuse that controls the circuit. The only tools you should need are a screwdriver and possibly a wire cutter. With the power turned off …

1. Remove the thermostat cover plate.

2. Unscrew the thermostat body.

3. Label all low-voltage wires with masking tape and then disconnect the wires.

4. Unscrew the thermostat base, taking care not to allow the wires to slip into the wall.

5. Pull the wires through the new thermostat base, secure the base, and then connect the wires to the appropriate terminals.

6. Make sure the wire connections at the transformer are secure. (The transformer will be in the same room as the heating/cooling unit or sometimes inside a furnace access panel.)

7. Install the required batteries in the thermostat body, and screw the body to the base.

8. Turn on the power at the service panel or fuse box, and test the thermostat.

Typical programmable thermostat.

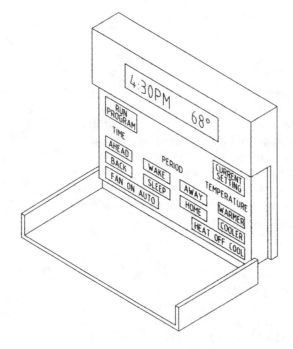

Wireless Thermostats

According to hydronic heating expert Joe Fiedrich (writing in *Contractor Magazine* in August of 1998), the first wireless, radio-controlled thermostat was introduced at a trade show in Frankfurt, Germany, in 1995 by the Velta Company. Like all wireless devices discussed in this chapter, a wireless thermostat saves you the trouble of wiring. The remote unit can travel with you as you roam around your house, enabling you to change the temperature at will. This could be a problem from an energy-saving standpoint if you keep changing the temperature every time you feel a bit cold or hot. It can really be a problem if several remote units are floating around with different family members.

Electrical Elaboration

Heating and cooling systems work best when you stick with a comfortable temperature range instead of changing the setting every time you feel a little too warm or too cool. Also keep in mind that you're never going to have a uniform temperature throughout your house. Walking around with a wireless thermostat in your pocket (or several units in the pockets of your spouse and kids) guarantees that you'll be eating up more energy as the setting keeps getting changed.

Wireless thermostat systems, now available on this side of the Atlantic, consist of a portable thermostat/controller (or several) and a receiver that mounts in place of your existing, wired thermostat. In new construction, the receiver is mounted near the furnace, thus voiding the need to run any wiring in the house. The system is programmable.

Recent Internet-advertised prices for these units have been around $250 for one portable thermostat and receiver and $125 for each additional thermostat. These prices are considerably higher than those for wired units.

The Wired World at Home

Current and future visions of a wired home life coincide, interestingly enough, with recent surveys that once again show America as the most overweight country in the world. Regardless of the validity of the surveys, it's apparent that we are becoming increasingly inactive, are eating more than we need, and are fueling the lazy life with wireless gadgets. Although I advocate installing plenty of wiring for low-voltage appliances and regular household current for lights, receptacles, and full-size appliances, I also ask you to think about just how automated you want your home life to be.

With enough of a budget, hardware, wire, and programming, your lights, music, bathwater, and eventually your refrigerator will do your remote bidding for you. Remote phones and intercoms give you hands-free contact with everyone from your pets to pizza-delivery services. If we chose to, we could spend even more time sitting than we do now. It's pretty tough to tell your kids to go outside and play when you're barely willing to get up and turn on a light because it's already programmed to turn on by itself.

My point is that, sure, you can install all sorts of wiring while your walls are open and your house is torn up a bit, but just consider *all* the costs of installing the devices, gadgets, and hardware before you start popping them in. Don't just think dollars, think common sense.

The Least You Need to Know

➤ An intercom can be a multifaceted helper that communicates, listens, and pipes music into your house.

➤ One advantage of an intercom is its capability to monitor visitors at your entry door.

➤ Door chimes are a simple, low-voltage installation, offering different tones for each entry door.

➤ When replacing a thermostat, make sure the new model is compatible with your heating/cooling system and is the same voltage as your old unit.

➤ When replacing low-voltage devices, make sure the old transformer is compatible with the new device.

➤ Wireless versions of hard-wired, low-voltage devices are available for cases in which installing wiring is either too disruptive or expensive.

Chapter 27

A Few Alternatives

In This Chapter

➤ Approaching home energy conservation

➤ Cutting down your electricity bill

➤ The cost of saving

➤ Alternative energy sources

As energy costs jumped in the 1970s and the number of users increased, the United States embraced conservation (although not to the extent that many would like). Insulation, insulated windows, and weather-stripping are all code-required in most residential construction. Fluorescent lighting is common in commercial buildings, as are sophisticated heating and cooling systems.

At the same time, however, we have more electrical toys, gadgets, and necessities than ever before. Utility companies are looking at deregulation, which will impact the fees we pay for electricity. With a booming economy, cheap oil, and sports utility vehicles and trucks outselling passenger vehicles, energy conservation isn't at the top of our list of important concerns *now*, but who knows what the future will bring? Will our usage today affect available power and its cost in the future?

Whether your concerns are financial or personal, there are plenty of ways to cut your energy consumption.

Conserve and Save

It's a pretty simple equation: Use less power, pay a lower utility bill. This doesn't mean you'll have to sit around in dark rooms wearing three layers of sweaters trying to ward off winter evening winds. You can do a lot by attending to the little things:

➤ Turn an electric baseboard down a few degrees.

➤ Use lower-watt lamps where appropriate.

➤ Run your washer, dryer, and dishwasher with full rather than partial loads.

➤ Turn off lights in unused rooms.

Many of these practices would barely cause a blip on the lifestyle scale, but every little bit adds up. You've seen the statistics: If every American household installed energy-efficient appliances, we would save enough electricity to power Portugal (or Aaron Spelling's Beverly Hills mega-mansion) for the next three centuries. In addition to saving us some cash every month, lower energy consumption means less need to build additional expensive power plants.

Electrical Elaboration

An energy audit will help you determine the best ways to reduce your energy consumption and, ultimately, your utility bills. An audit tests the efficiency of your heating and cooling systems, checks for leaks and drips, and snoops around for energy waste. You can do your own audit, or you can call a professional team that uses high-tech equipment to thoroughly test your home (for what you might view as a considerable fee). Some utility companies perform free or reduced-fee audits as well, so see if yours offers this service.

Watts Add Up, Doc

You can't avoid energy bills. Even if you live off 100-percent solar and wind power, you eventually have to buy replacement parts or bulbs for your light fixtures. We spend about a quarter of our electricity budget on lighting. This is more than $37 billion yearly.

We need light, but do we need as much as we use? Our grandparents (the ones who went through the Great Depression) were always turning off unneeded lights and admonishing our parents to do the same. A few years ago, I stayed at hotels in Europe with motion detectors in the hallways so that lights would be on only when someone walked by. We don't pay as much attention today because we pay without thinking much about it. But why throw the money away if we don't need to? Put another way, would you prefer to give an extra few hundred dollars a year to your local utility company or invest in Microsoft or a hot Internet company? When it's put that way, changing your lighting strategy seems like a good idea, doesn't it?

Energy-Efficient Lighting

We've already discussed the differences between incandescent and fluorescent fixtures and the advantage of the latter. Incandescent fixtures and lamps (a.k.a. bulbs) usually are the least expensive to buy, but they often are the most expensive to operate. You can realize both energy and cost savings by selecting the appropriate lamp and wattage for specific jobs.

One of the biggest improvements in lighting technology in recent years is the *compact fluorescent lamp* (CFL). It combines the efficiency of a fluorescent lamp with the convenience and size of an incandescent fixture. A CFL can replace an incandescent lamp that's about three to four times its wattage, saving up to 75 percent of the initial lighting energy. These savings come at a price, at least at first: A CFL costs 10 to 20 times more than a comparable incandescent lamp, but it will last 10 to 15 times as long. The Department of Energy calls CFLs one of the best energy-efficiency investments you can make.

Other basic energy-saving strategies for lighting include …

➤ Lowering the wattage of specific lamps or fixtures.

➤ Only using a light when it's needed instead of keeping it on when it's to no one's benefit.

➤ Using natural light whenever possible instead of electric lighting.

➤ Performing simple maintenance such as cleaning dust off fixtures and lamps once or twice a year.

➤ Replacing incandescent lamps with lower-wattage halogen lamps.

Bright Idea

You can cut energy consumption from lighting and improve light quality at the same time by following this simple rule: Establish your ambient lighting at the minimum acceptable levels while establishing your task lighting at the optimal level for the job at hand. In other words, use as little light as possible unless you're doing something that could be dangerous such as cutting food or wiring cable.

Heating

Programmable thermostats help regulate your furnace or baseboard heating, but regular upkeep is a must including ...

➤ Cleaning and replacing filters every three months in furnaces.

➤ Vacuuming dust from baseboard and in-wall heaters. (Be careful not to vacuum out the heating coils, however, because you can damage them.)

➤ Regular servicing of central heating systems by trained technicians.

Experiment with different temperature settings. You might find that you can be comfortable with a degree or two lower than your regular setting, and this translates into energy savings.

Thermostats

A thermostat is simply a temperature-sensitive switch that controls both heating and cooling. The least-expensive models are manually operated (you must change the settings by hand), but most people choose automatic or programmable thermostats. These allow a number of temperature settings throughout the day. You can lower the temperature, and thus energy consumption, while you're at work or asleep, and you can raise the temperature while you're up and around.

Over the years, there has been some dispute regarding fuel savings from using variable settings (a.k.a. temperature setbacks). According to the Department of Energy, however:

"A common misconception associated with thermostats is that a furnace works harder than normal to warm the space back to a comfortable temperature after the thermostat has been set back, resulting in little or no savings. This misconception has been dispelled by years of research and numerous studies. The fuel required to reheat a building to a comfortable temperature is roughly equal to the fuel saved as the building drops to the lower temperature. You save fuel between the time that the temperature stabilizes at the lower level and the next time heat is needed. So the longer your house remains at the lower temperature, the more energy you save.

Another misconception is that the higher you raise a thermostat, the more heat the furnace will put out, or that the house will warm up faster if the thermostat is raised higher. Furnaces put out the same amount of heat no matter how high the thermostat is set—the variable is how long it must stay on to reach the set temperature. In the winter, significant savings can be obtained by manually or automatically reducing your thermostat's temperature setting for as little as four hours per day. By turning your thermostat back 10 to 15 degrees for eight hours, you can save about 5 to 15 percent a year on your heating bill—a savings of as much as 1 percent for each degree if the setback period is eight hours long. The percentage of savings from setback is greater for buildings in milder climates than for those in more severe climates. In the summer, you can achieve similar savings by keeping the indoor temperature a bit higher when you're away than you do when you're at home."

A programmable thermostat will store and repeat more than half a dozen temperature settings each day, any of which can be overridden manually if necessary. Honeywell makes an excellent, popular line of these thermostats starting at around $75.

Electrical Elaboration

Programmable thermostats come in a variety of types and use various technologies. You want a thermostat that's compatible with the electrical wiring connected to your current thermostat. Choose a model with a battery-operated backup so you don't lose your settings if the power goes out. Clear instructions are a must, especially if they are printed on the cover of the thermostat for quick reference. Only a few programmable thermostats are available for line voltage control of baseboard heaters, so be sure your supplier carries these models.

A programmable thermostat is more dependable than human control of your household temperature because the latter usually comes down to an unreliable litany of "It's freezing in here," "I'm boiling," and "Who messed with the thermostat?"

Air-Conditioning Alternatives

The Department of Energy is a wonderful resource for energy-saving ideas, and I have plundered and quoted it shamelessly for this chapter. Nevertheless, I doubt its recommendations for natural house cooling will go over well in, say, west Texas.

Look at the problem of cooling your house from two perspectives:

➤ You need to cool and dehumidify the interior of your home.

➤ You need to prevent heat buildup in the first place.

In the summer, heat is absorbed into your house. Your walls, roof, and windows soak up the sun. Large appliances such as ovens and dryers also add to the problem. A strategy to deflect outside heat will go a long way toward cooling your house. Some strategies include …

➤ Painting your house a light color, which reflects heat away, rather than a heat-absorbing dark color.

➤ Applying reflective coatings to window glass. (These will cut down on light transmission, a drawback during the winter. It's also best if these film-like coatings are applied by a professional.)

➤ General weatherization and insulation, especially attic insulation.

➤ Landscape shading (such as trees), which can reduce indoor temperatures by as much as 20°F.

➤ Using window shades, awnings, and shutters.

➤ Natural ventilation using the coolest or breeziest parts of the day, coupled with cross-ventilation as well as attic ventilation to reduce accumulated heat.

➤ Using window and ceiling fans.

➤ Using heat-generating appliances during the coolest part of the day or evening.

Improving Your Air Conditioning

Air conditioners are made up of an evaporator, a condenser, copper tubing, a compressor, and refrigerant. You'll either have individual room air conditioners or central air conditioning. With regular maintenance, an air conditioner can last for many years. Problems, and therefore poor energy usage, can result from the following:

➤ Improper operation (such as windows or outside doors left open while the air conditioner is on)

➤ Poor installation (leaky ducts, poor fitting of window units)

➤ Improper charging of refrigerant by technicians or even leaking refrigerant

➤ Dirty filters or dirty air-conditioning coils

Regular air-conditioner maintenance is explained in your owner's manual. Even something as simple as cleaning or replacing the filters will do wonders for efficiency. Sealing ducts is as important for air conditioning as it is for heating. According to the Department of Energy, studies indicate that "10 percent to 30 percent of the conditioned air in an average central-air-conditioning system escapes from the ducts." A yearly checkup by a competent technician will keep your system working safely and efficiently. Your technician should …

➤ Check the amount of refrigerant.

➤ Check for leaks.

364

➤ Inspect the ductwork for leaks.

➤ Measure airflow through the evaporator coil.

➤ Check all electrical connections.

➤ Oil motors and check the belts.

➤ Check the accuracy of the thermostat.

The ultimate improvement is replacing an old system with a more efficient new one. Some of the best new units use 30 to 50 percent less energy than air conditioners from the mid 1970s and as much as 20 to 40 percent less than air conditioners that are 10 years old.

Sizing Up Your Air Conditioner

Bigger isn't better when it comes to air conditioners. A reputable dealer will use one of the calculation formulas published by the Air Conditioning Contractors of America (ACCA) or the American Society of Heating, Refrigerating, and Air Conditioning Engineers (ASHRAE) to correctly size your central-air-conditioning system. A system that's too big will be more expensive than the system you should have, it will cost more to run, and because it will cycle on and off more frequently, your rooms will be less comfortable and more humid. On top of that, the frequent cycling will wear out the compressor sooner than normal.

Time to Get Efficient

Putting our tax dollars to work, I'll let the Department of Energy explain it best:

"Each air conditioner has an energy-efficiency rating that lists how many BTU per hour are removed for each watt of power it draws. For room air conditioners, this efficiency rating is the Energy Efficiency Ratio, or EER. For central air conditioners, it is the Seasonal Energy Efficiency Ratio, or SEER. These ratings are posted on an Energy Guide Label, which must be conspicuously attached to all new air conditioners. In general, new air conditioners with higher EERs or SEERs sport higher price tags. However, the higher initial cost of an energy-efficient model will be repaid to you several times during its life span. Your utility company may encourage the purchase of a more efficient air conditioner by rebating some or all of the price difference. Buy the most efficient air conditioner you can afford, especially if you use (or think you will use) an air conditioner frequently and/or if your electricity rates are high."

Cool Your Hot-Water Costs

The Department of Energy figures that one seventh of your electric bill goes toward heating your water if you use an electric water heater (and have all electric, not gas, appliances). Let's face it, we like hot showers, and we should—they're one of modern life's great pleasures. Still, if we can reduce our energy costs without noticeably reducing our pleasures, why not see what it takes?

There are two basic approaches: Reduce the amount of hot water you need, or improve your means of heating it. Hot water finds four main endpoints in our homes:

➤ Faucets

➤ Showers/tubs

➤ Dishwashers

➤ Washing machines

Some simple changes and maintenance can cut down your water usage and your energy costs.

Drips and Flows

We've all ignored them at one time or another—those pesky drips from sink or tub faucets. A leak of one drip per second can cost $1 per month, but it can be repaired in a few minutes. Installing low-flow showerheads and faucet aerators will always cut down on water consumption but still provide plenty of water flow. I'll be the first to admit there's nothing quite like a large, old showerhead that lets loose with gallons and gallons of water, but these just won't cut it in the looming future of water restrictions, especially in the western states. One caveat: Buy a quality low-flow showerhead; the cheap ones will be disappointing and can drive you back to your old water-using ways.

Aerators vary as to how much water flow they allow. You'll probably want a higher flow in your kitchen sink than in your bathroom.

The Appliances

Dishwashers are an energy-efficient means of washing your dishes when used for full loads. You also can cut back on your energy usage by skipping the dry cycle. Many new dishwashers come with multiple settings that enable you to choose a shorter wash-and-rinse cycle when appropriate. A new dishwasher also will be more energy efficient than older models. Newer washing machines enable you to choose water temperatures and water levels for different-size loads (although washing full loads is an energy-saving strategy).

Your Hot-Water Tank

Turning down the temperature on your hot-water tank not only will reduce energy consumption, it also will reduce the chance of being scalded. A tank set at 120°F will suit most household requirements. For each 10°F reduction in water temperature, water-heating energy consumption can be reduced 3 to 5 percent. Many jurisdictions require water-heater installers to set the temperature at 120°F.

Do you have accessible hot-water pipes running through your basement or crawl space? Pipe insulation, available at hardware stores and home centers, can easily be installed by most homeowners and reduces heat loss. Insulation blankets reduce heat loss from your hot-water tank, especially older models. (Some new ones don't need additional insulation.) Check with your local utility, which might offer blankets at lower costs. A properly installed insulation blanket should pay for itself within one year through realized energy savings. If your water heater is more than 10 years old, consider replacing it with an energy-efficient model.

Bright Idea

Try cooking an extra hot dish or two the next time you're baking something in your oven. You'll use the same amount of heat to cook one pan of lasagna as to cook two or three. You also won't need as much energy to warm it up when you're ready for a future meal as you would to cook it from scratch.

Keeping the Outdoors Outdoors

Once again, here is a quote from the Department of Energy:

> "Caulking and weatherstripping are the easiest and least expensive weatherization measures and can save more than 10 percent on energy bills. Caulking and weatherstripping are most often applied to doors and windows, which account for about 33 percent of a home's total heat loss. Insulation is probably the most important consideration in improving the energy efficiency of a home."

There's no shortage of caulking and weatherization materials at hardware stores and home centers everywhere. Your choice of materials will depend on the age and fit of your windows and doors and the material you're caulking. Look at all materials carefully, and install your selected product on one window or door before committing yourself to a shopping bag full of the stuff. Again, your local utility probably will have information and pamphlets about effective weatherization for your part of the country.

Look Through Any Window

Old single-pane windows, especially metal windows, are less efficient than newer insulated units in terms of heat loss. Lost heat means your furnace or zone heaters have to kick in that much harder to keep your home warm. Short of replacing all your windows (an expensive proposition, especially if you have old, ornate, wood windows), you can improve your current windows. Improvements include …

➤ Weatherstripping.

➤ Interior or exterior storm windows.

➤ Caulking.

Storm windows range from sheets of plastic adhering to your interior trim to custom wood-and-glass exterior units. They all have their drawbacks in appearance, durability, or cost, but they cut down on heat loss. New windows will, too, but they'll take years to pay back their cost in energy savings.

Insulation

In the average American home, heating and cooling soak up between 50 and 70 percent of the energy consumed. Insulation not only cuts your energy costs, it also helps maintain a more uniform temperature year-round and adds sound-dampening properties as well. The following is the usual recommended order, from most important to least, for installing insulation:

➤ Attic

➤ Floors above unheated spaces

➤ Walls in heated basements or unventilated crawl spaces

➤ Exterior walls

Recommended levels of insulation vary depending on where you live. Attics are critical, especially in cold climates. Your heat, which rises throughout your house, has to go somewhere, and flowing through your attic is its last logical choice. Insulation greatly reduces this loss.

Here Comes the Sun

Not exactly a winning strategy in the rainy state of Washington, solar heating and power generation probably will get more exposure in the future as the technology improves. Residential solar-heating collectors absorb sunlight, heat up either air or a liquid, and convert it into heat energy. These collectors include …

➤ Flat-plate collectors.

➤ Evacuated-tube collectors.

➤ Concentrating collectors.

368

Solar power and heating are beyond the scope of this book. This is an evolving energy strategy—one that is more common is some parts of the country (the Southwest or Florida, for example) but by no means prevalent. Your savings will depend on the local cost of other power sources (electricity, gas, and oil) and their long-term outlook. There also is the installation cost of solar versus conventional systems.

Nevertheless, solar heating can be effective in some situations. I would do plenty of research before you start tearing out your water heater and furnace and telling your neighbors you're no longer going to be a slave to the power companies after you install your rooftop solar collectors. They might remember your gloating the first time you knock on their door and ask to use the shower because your tap water is two degrees above freezing.

Is It Worth It?

It takes money to save money (and, in this case, energy). Some of the strategies mentioned in this chapter are simple lifestyle changes that cost nothing; others, such as upgrading appliances or adding insulation, have a price tag attached. How do you know if the savings are worth the investment?

The Department of Energy puts it this way:

Positively Shocking

In almost all climates, you'll need a conventional hot-water tank as a backup system to solar water heaters. Many building codes actually require this if you go solar. Also, local zoning or covenant restrictions might dictate where you can place your collectors and what size they can be.

> "Generally, there are two ways to analyze the costs of energy-efficiency investments: the simple payback period, which is the amount of time required for the investment to pay for itself in energy savings; and the full life-cycle cost, which is the total of all costs and benefits associated with an investment during its estimated lifetime.

You can obtain an estimate of the simple payback period by dividing the total cost of the product by the yearly energy savings. For example, an energy-efficient dryer that costs $500 and saves $100 per year in energy costs has a simple payback period of five years. Computing life-cycle costs is more difficult. Life-cycle costing is a method of economic evaluation in which all values are expressed as present dollars. This evaluation method sums the discounted investment costs (less salvage value); the operation, maintenance (nonfuel), and repair costs; the replacement costs; and the energy costs of an appliance or building system. For definitions of these terms and the formula for performing life-cycle cost analyses, see the "Life-Cycle Costing Manual for the Federal Energy Management Program," NBS Handbook 135, Revised 1987. This manual is available from the Superintendent of Documents, U.S. Government Printing Office, Washington, DC, 20402.

Before making your decision, examine your budget, the expected payback periods, and the estimated lives of different alternatives. Products or systems with payback periods that approach or exceed their projected life usually are not worthwhile. Compare the life-cycle costs of similar products or systems. These include installation (if any), operation, and maintenance costs.

Will you retrieve the cost of the improvements such as a new appliance? Are there other values to consider such as your desire to simply save energy rather than money? And what about disposal cost in a landfill of your old appliances and so on?"

There are other comfort and even aesthetic considerations. Do you want to wear sweaters indoors all winter because you've dropped the temperature down to 67 degrees? Do you really want to look at sheets of plastic spread out over your 1920 bungalow windows? If you live alone, do you really want to wait until you have a full laundry or dishwasher load before you run your appliances? If you rarely cook and live on tech food—soda, pizza, take-out Chinese—it could be a month before you run the dishwasher. Anything sitting in there that long is going to be a little crusty by then.

As you can see, you have a number of issues to consider while you're trying to cut down on your energy usage and cost.

Resources

Air Conditioning and Refrigeration Institute
4301 North Fairfax Dr., Suite 425
Arlington, VA 22203
Phone: 703-524-8800
Fax: 703-528-3816
E-mail: ari@dgsys.com

American Council for an Energy-Efficient Economy (ACEEE)
1001 Connecticut Ave., Suite 801
Washington, DC 20036
Research and Conferences: 202-429-8873
Publications: 202-429-0063

American Solar Energy Society (ASES)
2400 Central Ave., Unit G-1
Boulder, CO 80301
Phone: 303-443-3130
Fax: 303-443-3212

Association of Home Appliance Manufacturers
20 N. Wacker Dr., Suite 1231
Chicago, IL 60606
Phone: 312-984-5800
Fax: 312-984-5823

Florida Solar Energy Center (FSEC)
1679 Clearlake Rd.
Cocoa, FL 32922-5703
Phone: 407-638-1015
Fax: 407-638-1010
E-mail: webmaster@fsec.ucf.edu

Honeywell Inc.
Honeywell Plaza
P.O. Box 524
Minneapolis, MN 55440
Phone: 612-951-1000
Customer Response Center: 1-800-345-6770

National Association of State Energy Officials (NASEO)
1414 Prince St., Suite 200
Alexandria, VA 22314
Phone: 703-299-8800
Fax: 703-299-6208
E-mail: info@naseo.org

National Insulation Association
99 Canal Center Plaza, Suite 222
Alexandria, VA 22314-1538
Phone: 703-683-6422
Fax: 703-549-4838

National Wood Window and Door Association
1400 East Touhy Ave., Suite 470
Des Plaines, IL 60018
Phone: 1-800-223-2301
Fax: 847-299-1286

U.S. Department of Energy's Energy Efficiency and Renewable Energy Clearinghouse (EREC)
P.O. Box 3048
Merrifield, VA 22116
Toll-free: 1-800-DOE-EREC (1-800-363-3732)
TDD: 1-800-273-2957
BBS: 1-800-273-2955
Fax: 703-893-0400
E-mail: doe.erec@nciinc.com

U.S. Department of Energy's Office of Building Technology State and Community Programs (BTS EE41)
1000 Independence Ave. SW
Washington, D.C., 20585

Of course, things change, so use your favorite Internet search engine to supplement this list.

The Least You Need to Know

➤ Many home energy-saving strategies involve only minor work on your part.

➤ Investments in energy conservation will more than pay for themselves, sometimes surprisingly quickly.

➤ It only takes a few minutes each day (a shorter shower, turning a light off when it's not needed) to start your own conservation program.

➤ Careful consideration of payback periods will guide you to the best energy-saving decisions.

Putting It All Together

By now, you've familiarized yourself with your electrical system and any of its short-comings. You've paged through this book and have gotten some notion of how much work is necessary to upgrade or modify this system. Maybe you've been sketching out some locations for new yard lights or automated features you'd like to install. You just need to tally up the cost in both time and dollars to create this electrical wonderland.

Your choices aren't exactly limitless, but they certainly are extensive. I can't say we're the most gadget-oriented people in the world (the Japanese have a lot of fun toys, too), but we've never met an electrical or mechanical convenience we didn't like. You'll pay a price for whatever degree of convenience and sense of fashion and design you desire. A basic porcelain light fixture with a bare bulb will do the job of providing you with safe, effective lighting, but who wants to look at that in a kitchen or a bathroom? No one needs a food disposer, but many people won't buy a new house without one. Every additional upgrade or new appliance adds to the cost of your electrical system.

Money aside, much of what you're buying is simply comfort, safety, and accommodation. A heat lamp in a bathroom, a programmable thermostat, and an intercom system all make life easier. An outdoor motion detector adds a sense of safety and provides lighting at night. Something as simple as adding circuits and receptacles in an old house can improve day-to-day tasks.

Bright Idea

One of the simplest ways to choose improvements for your old electrical system is to walk through some new homes during weekend open houses. Look carefully at the number of receptacles and lights, both inside and outside the house. Note any low–voltage wiring as well.

How far you take your upgrades and improvements (and which ones you do yourself versus hiring out) is strictly a personal decision. No judge on a Municipal Court of Electrical Affairs is going to issue you a summons demanding that you raise the quality of your electrical system (although a bank might require a new service panel before granting a mortgage in some cases). Consider your needs, safety requirements, and budget and go from there.

How Far Do You Go?

Fuse systems without grounding, combined with knob-and-tube wiring, are not unsafe unless they are asked to do more than they're designed to do. You must monitor your usage and live with certain restraints. If you find this too restrictive, your first order of business is to replace your fuse box with a new, up-to-code service panel with circuit breakers. A grounded, 200-amp service panel is more or less the bottom line for admission into the world of electrical civilization.

After you've gotten your new service, you can decide how much to upgrade. Circuits for receptacles usually are easier to add or upgrade than those for ceiling lights. Why? Because usually you can access the space between wall studs by drilling through from an open basement or attic area, but drilling through a ceiling joist can require a lot of plaster or drywall repair. The cost of wall and ceiling repair and painting just adds to your installation costs.

You should at least add the following circuits:

➤ A new laundry circuit

➤ Dedicated GFCI kitchen circuits

➤ GFCIs in the bathrooms and outdoors

The next step is running branch circuits for additional receptacles, improving existing lighting, and running dedicated circuits for offices, workrooms, and garages. In a house with a new or newer electrical system, you'll probably add more phone lines, phone jacks, coaxial cable, and low-voltage wiring. Your job with any of these improvements is to pick and choose among them with a clear understanding of what is required to complete each project.

Time and Money Considerations

Money is a convenient measuring tool. It takes you so many hours or days to earn so many dollars. After taxes and other mandatory expenses (mortgage, utilities, food), whatever is left over can go to your electrical improvements. You also can finance the improvements, but this still requires money to pay back the loan. Either way, it's going to cost something. You have to decide whether the dollars spent will be worth the results. Remember, you're paying in after-tax dollars. (When estimating the true cost of work, remember that since you pay taxes, you end up with less money in your pocket than you earned in salary, because some of that salary goes to the government.)

Time is another important consideration. If you have plenty of time available to work on your house, you can put this time to use profitably. You can rent or buy the same drills that an electrician uses and drill through floor joist and wall studs with equal aplomb. Drilling is not computer science. This is a good thing because if your drill was a computer, it probably would stop inexplicably, demand to be restarted from time to time, and end up being incompatible with certain drill bits. During remodeling, much of an electrician's time is spent simply accessing walls and ceilings and pulling cable. You might not want to do the final connections, but you certainly can do much of the work up to that point.

That said, paying an electrician to do the work might appear to be more expensive than doing the work yourself, but this isn't always the case. It might be cheaper for you to work overtime or to do a freelance job and use those earnings to hire an electrician. You might prefer writing a computer program to crawling around your attic pulling cable. At these times, "Know thyself" is good advice.

Positively Shocking

If you borrow money to pay for electrical upgrades, you have to figure the complete cost of the loan to calculate any increased value in the house versus the cost of the improvements. The longer you take to repay the loan, the less economical the improvements, although they do add to the value of your home.

Added Value

At its most basic, an improved electrical system brings more value to your home. You won't necessarily increase the value of your house, however, by the same amount you spent improving it. The only way to know for sure is to calculate the following:

➤ The current value of your home

➤ The cost of the improvements

➤ The projected value of your home after the improvements are made

Positively Shocking

Before trading work time with a friend for your electrical jobs, be sure such work is legal where you live. In the state of Washington, for example, it is illegal to perform electrical work at a property you do not own unless you are a part of the immediate family. Friends do not count.

It's tough to figure the value of electrical upgrades. Kitchen and bathroom improvements, for example, often are cited as two remodeling projects that return most or all of their costs. You could argue that any electrical improvements done during these projects will recover most or all of their cost, but there's no accurate way to separate them from the job's total outlay. A new service panel might make your house easier to sell, and this in itself might save you some money if you're trying to get out from under a higher-rate mortgage or if you've already purchased another property. It's safe to say you'll recover some of the cost of your electrical upgrades and changes, but it's difficult to figure out how much you'll recover.

Convenience

How do you put a price on the convenience of an overhead light where you want it or a dedicated circuit for your computer? This is strictly subjective, but few people would complain about having too many receptacles or lights, even if they don't use them all. In truth, some receptacles don't get used very often, but we want them when we want them. It's better to have too many than to have too few.

Bright Idea

Insurance companies like new electrical systems, especially circuit breakers. Some companies won't offer insurance for homes with knob-and-tube wiring and fuses. Those that do might offer you a reduced rate after you upgrade.

Estimating the Job

As a do-it-yourselfer, you can pretty easily add up the cost of cable and all the components needed for your electrical improvements. Just make a list and head to your nearest mega-size home-improvement store. It's not a bad idea to start there so you can look at different box sizes and fixtures. Make a list that will include, depending on your project ...

➤ Cable, noting different gauges.

➤ Switches and receptacles.

➤ Electrical boxes, all types. (Remember, buy the biggest ones your wall can accommodate.)

➤ Cover plates.

➤ Circuit breakers.

➤ Light fixtures (brand names only, please).

➤ Appliances (disposer, dishwasher, and so on).

➤ Miscellaneous items such as wire nuts, electrical tape, and light bulbs.

The easiest way to buy cable such as 12/2 AWG is by the 250-foot box. You might need less, but it's always good to have cable around for future projects, and it's not very expensive. If you need to buy or rent tools, these will have to be added to your budget as well. Remember, you want to rent tools for as little time as possible to keep your costs down. Be sure you have all your walls and floors open so you can do all your drilling at one time.

Electrical Elaboration

Large electrical jobs (such as major rewiring) incur the inevitable fast-food costs. If you're working evenings and weekends or during your vacation, the last thing you want to do after drilling, pulling, and tearing holes in your walls is cook a meal. Tempers also can flare if you and your spouse are doing the work and are trying to communicate by yelling from one floor to another. Know when to put the tools down and get away from the job. There's nothing admirable about toughing it out when you're tired, hungry, and angry.

Boxes, devices, and cable are comparatively low-priced items. You won't find one receptacle costing five times as much as a comparable unit from another manufacturer. Fixtures, on the other hand, will be the real budget buster. The aforementioned basic porcelain ceiling light only costs a few dollars, but a chandelier can cost thousands. They both supply light, but a bare-bulb fixture isn't what most people want to look at in a formal dining room.

This is where it pays to do plenty of shopping around, comparing styles and prices. Ask your local lighting stores when they have their yearly sales and what kind of discounts you can expect. Look for one-of-a-kind items, fixtures missing their original packaging, or globes being sold at a reduced price. If you're willing to consider a different fixture than the one you originally had in mind, you might find a real bargain. That said, a cheap fixture—one with a bad wiring assembly or glass globe—is no bargain. Make sure there are no restrictions listed on the fixture (for example, being listed for nonresidential use only). If one fixture costs considerably less than a brand-name counterpart, you might have problems.

Positively Shocking

In your pursuit of fixtures and devices, be sure to check for a UL tag or a tag from a similarly approved national testing organization. It's not an assurance of quality *per se,* but it does ensure that the product passed minimum testing.

A Realistic Time Frame

The amount of time it will take you to complete an electrical job or repair will depend, to some extent, on how much time you have available. The actual number of hours required to run a circuit will be the same whether you do it an hour a night for a week or spend a whole Saturday doing the job. For some jobs, stretching out the completion date won't make any difference (adding extra circuits, for example). Gutting your bathroom wiring is another matter. Do you want to do your grooming chores by flashlight because you haven't finished putting in the new lights yet?

Some people believe you should take any remodeling estimate and double the time (and sometimes double the money) if you want a more realistic figure. This is too simplistic, but adding a 25 or even 50 percent "fudge factor" isn't a bad idea. After you've done a few electrical jobs, you'll have a better idea about your abilities and the speed with which you can complete the work.

Sample Jobs

Each house is different. Access from your service panel to the job site (the wall or ceiling where you're doing an installation) requires cutting through wood and plaster or drywall to run your electrical cable or wiring. Some homes, such as adobe or certain masonry structures, pose unique challenges because they do not have typical 2×4 studs in the walls. You'll probably be dealing with standard wood framing, however.

An unfinished basement and attic can really speed up your job. They enable you to drill and pull cable through open joist and wall plates, giving you access to the cavity between the wall studs on the first floor of your house. Access to the second floor requires cutting into your first-floor ceiling. You can limit this cutting to a single location (if it works out for your wiring plans) by running multiple cables up through one pair of wall studs.

Electrical Elaboration

Adobe housing is built from clay bricks instead of standard wood framing. A pathway is routed out of the bricks to accommodate electrical conduit, which houses the electrical cable. The conduit is then covered with plaster. Some adobe homes, however, have wood framing added to the interiors to facilitate wiring and plumbing (it's easier to cut up wood to add pipes and wires than it is to cut into clay).

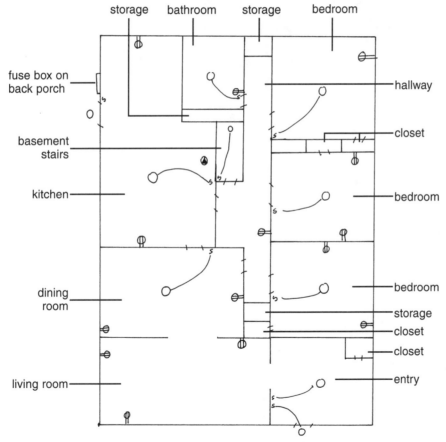

1924 bungalow before. All receptacles and lights were on old, two-wire system without a grounding conductor.

GFCI

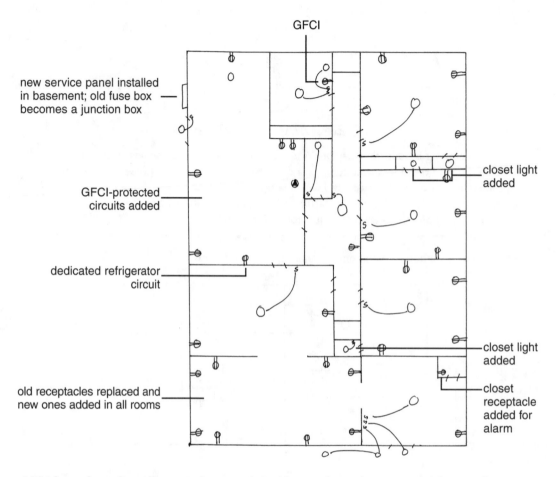

new service panel installed
in basement; old fuse box
becomes a junction box

GFCI-protected
circuits added

dedicated refrigerator
circuit

old receptacles replaced and
new ones added in all rooms

closet light
added

closet light
added

closet
receptacle
added for
alarm

1924 bungalow after. All receptacles upgraded with new three-wire system (with grounding conductor).

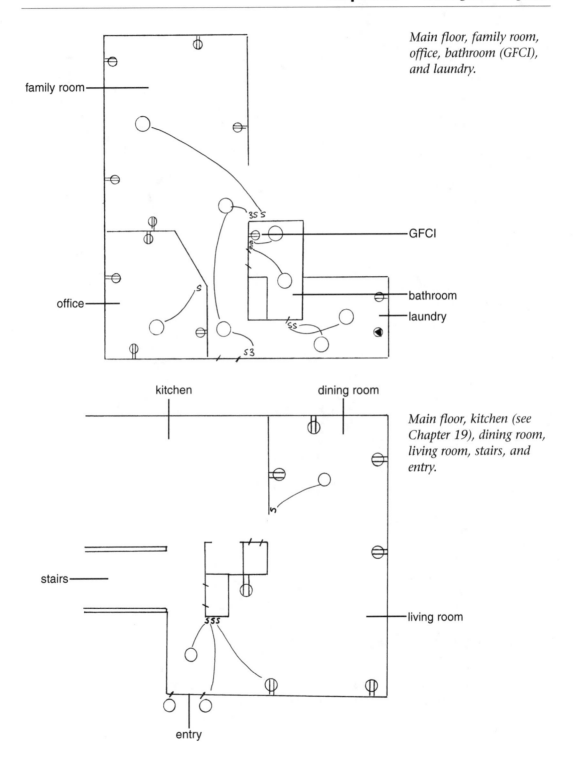

Main floor, family room, office, bathroom (GFCI), and laundry.

family room

office

GFCI

bathroom

laundry

kitchen

dining room

Main floor, kitchen (see Chapter 19), dining room, living room, stairs, and entry.

stairs

living room

entry

Second floor.

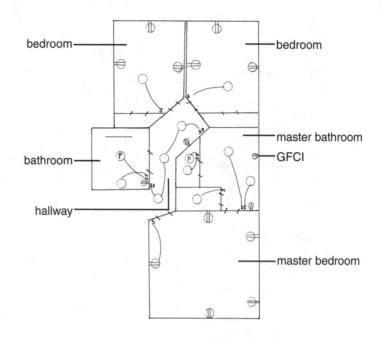

Garage.

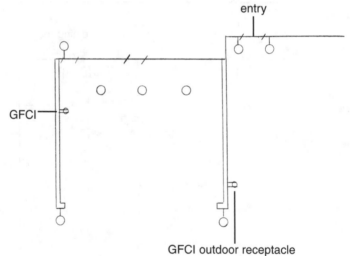

Hiring an Electrician

We don't all do all tasks equally well. I wallpapered a room once, and that was enough to tell me never to try it again. A qualified electrician (not a "handyman" who does a little of everything) should do your job neatly, in good time, and in accordance with your local code. You can always cut your costs by drilling holes and pulling cable yourself, as I've emphasized several times in this book, but if free time is elusive and you're a two-income household, hiring an electrician might be a better way to go. Check your local code regarding this "co-work" kind of arrangement. You'll probably need to obtain a permit for your portion of the work while the electrician obtains another permit for the rest.

With the knowledge you've gained in this book, you'll be able to intelligently discuss any potential work with an electrician and understand your options. Just be sure that you do the following:

➤ Get a written estimate.

➤ Schedule a time to start and complete the work.

➤ Confirm that the electrician is licensed and bonded.

➤ Make sure the work is done with a permit and is inspected.

➤ Pay your electrician's bill in a timely manner.

Positively Shocking

Taking out a permit isn't enough. You must have an inspector sign off on the work. A permit can be taken out and allowed to expire without the work ever being inspected!

The Least You Need to Know

➤ There usually are more advantages to upgrading an old electrical system than savings in not upgrading it.

➤ Look at your costs from all angles: pretax and after-tax dollars, the value of your time, and your willingness to install less-expensive fixtures than originally planned.

➤ Upgrades will add some value to your home but not necessarily enough to equal the money you paid for them.

➤ You can't measure increased convenience and safety in dollars.

Resources

Plenty of books on electrical repair are available, as well as remodeling and home-repair texts that have sections devoted to electrical work. The Internet also offers an abundance of information, but watch out for independent sites that may not be accurate. Everyone has an opinion, but the only one that will ultimately matter is your electrical inspector's. Sites associated with known electrical organizations or remodelers (such as *Hometime*) are a safe bet.

Web sites come and go and books become outdated. (Remember, the NEC is reissued every three years.) Most wiring techniques don't change drastically, but you must be aware of code changes, specifically on a local level because those are the ones that will affect you. Nevertheless, the following books and Internet information will further assist you in your rewiring projects.

Books

Black & Decker's The Complete Guide to Home Wiring: A Comprehensive Manual from Basic Repairs to Advanced Projects. Minnetonka, Minnesota: Cowles Creative Publishing, 1998.

This book is loaded with color photographs showing everything from examples of old wire to information about how to install outdoor lights. It is clear and well-laid-out.

Cauldwell, Rex. *Wiring a House*. Newtown, Connecticut: Taunton Press, Inc., 1996.

Rex Cauldwell is a master electrician with more than 30 years of experience and opinions. He explains how to choose tools and materials, design the job, and do safe installations. Note: This book was written according to the 1996 National Electrical Code.

Richter, H. P. *Wiring Simplified: Based on the 1999 National Electrical Code*. Revised by W. Creighton Schwan. Somerset, Wisconsin: Park Publishing, 1999.

This book is in its thirty-ninth edition, which should tell you something. The publisher says, "This long-time favorite is packed with information presented in clear language." The compact book fits easily into a toolbox and features 225 illustrations.

Web Sites

www.bbb.org
The Better Business Bureau's site provides information about hiring a contractor and contacting your local office to check out his or her business record.

www.codecheck.com
At this site, you can look up the history of knob-and-tube wiring as well as code issues and links to other electrical sites.

www.faqs.org/faqs/electrical-wiring/
This site answers many common electrical questions.

homerepair.miningco.com/msubelec.htm?iam=mt
This site discusses common electrical repairs.

www.hometime.com
This remodeling site's amiable hosts also have their own television show (called *Hometime,* naturally enough). The site is down-to-earth with easily understood explanations.

www.hut.fi/Misc/Electronics/wiring.html
This site contains plenty of information and links.

www.inspect-ny.com/aluminum.htm
Information about residential aluminum wiring is provided at this site.

www.kitchenenet.com
This site offers manufacturer links and free information about the kitchen and bath industry.

www.nfpa.org
This is the site for the National Fire Protection Association, which publishes the National Electrical Code.

peripherals.miningco.com/msub_power.htm?iam=mt
This site discusses power protection and surge suppression for your computer and peripherals.

www.pueblo.gsa.gov/press/electric.htm
The government says it best: "Faulty electrical equipment and misuse of electrical appliances are frequent causes of accidents and fires. To help you identify potential trouble spots and reduce the risk of electrical accidents in your home, the National Electrical Safety Foundation in cooperation with the U.S. Consumer Product Safety Commission has published a new booklet titled *A Home Electrical Safety Check.*"

www.soundhome.com
This well-received site offers remodeling and home-inspection information.

www.ul.com
This is the Web site of the Underwriters Laboratory.

Remember, the Web changes constantly. Use Internet search engines to augment this list.

Glossary

alternating current (AC)   A flow of electrons that increases to a maximum speed in one direction (toward a load in a residential electrical system), decreases to zero, and then returns to its source (your service panel), reaching maximum speed in this reversed direction as well. These cycles are repeated again and again at the same intervals of time: 60 cycles per second in the United States and 50 cycles per second in Europe.

amp or **ampere**   A unit of measurement for electrical current, specifically the number of electrons (coulombs per second) passing a single point in a conductor in one second (the "'volume'" of electricity).

ampacity   The amount of current, in amperes, that a conductor can carry without exceeding its temperature rating.

apprentice   An electrician in training.

armored cable   Electrical conductors protected by a metal sheathing. BX is a common form of armored cable.

AWG   American Wire Gauge, a standard for wire size.

branch circuits   The circuit conductors running between a fuse or circuit breaker and the outlet(s).

breaker   An overcurrent safety device that protects a circuit. It also is called a circuit breaker.

bus bar or **bus**   Metal strips in a service panel. Circuit breakers are connected to a hot bus, and the neutral and grounding conductors are secured in the neutral and ground buses.

BX cable   A form of armored cable that is no longer used.

cable clamps   Clips that secure electrical cable to an electrical box.

certified journeyman   In the electrical trade, one who has passed through an apprenticeship and is qualified to work independently on electrical installations and repair.

circuit   A current of electricity that originates at a service panel or fuse box and is carried by a hot conductor to a device. A neutral wire provides a path for the return current, thus forming a complete loop. The complete loop is the circuit.

conduit   Metal or plastic tubes for protecting individual wires.

Coulomb   The number of electrons in a one-amp current that pass through an individual point on a conductor in one second.

cover plate   A plate that protects a device and the contents of an electrical box.

damp-location fixture   A fixture designed for wet locations, such as above a shower, that provides some protection against dangerous electricity.

dedicated circuit   A circuit limited to supplying a current to a single load or to limited load(s).

device   A piece of hardware in an electrical system that carries a current but does not use it. Examples include a switch or a receptacle.

direct current   An electrical current that flows in only one direction.

drip loop   The hanging position of an incoming service drop to a service head on the roof of a house; the loop prevents water from entering the service head.

duplex receptacle   A receptacle that can hold two plugs.

electrical box   A protective plastic or metal box that houses wire connections and devices.

fixture   Generally, any lighting device such as a table lamp, floor lamp, sconce, recessed light, and so on.

fluorescent lighting   A major form of lighting that is an alternative to incandescent lighting.

fuses   Replaceable overcurrent devices that have been replaced in modern housing with circuit breakers. Fuses commonly are used for motor loads such as AC, HVAC, or pool equipment.

ground   A large conducting body (such as the earth) used as a common return for an electric circuit due to its zero potential; the endpoint of a current.

ground fault   Misdirected current from the hot wire or neutral wire to a ground wire, an electrical box, or another conductor; an undesirable current path from a conductor carrying a current to ground.

ground-fault circuit interrupter (GFCI)   A GFCI, in the form of a circuit breaker or a receptacle, detects misdirected current and shuts it off faster than a standard circuit breaker can. This is useful in places such as bathrooms where misdirected current is more likely.

grounded   The term used to describe an electrical system connected to the earth.

grounded conductor   The white or neutral conductor that is intentionally grounded.

grounding conductor   The bare copper or green insulated conductor that connects the electrical system or equipment to the earth by being bonded to a water pipe and/or a grounding electrode.

hot   Term used to describe a conductor that carries a current from a source (ultimately, the service panel). A nongrounded, current-carrying conductor.

hot bus bar   A metal strip that carries incoming current through a service panel.

incandescent lighting   A resistance form of lighting.

joist   A horizontal, wood framing component of a floor.

kilowatt hour   Kilowatts multiplied by hours. This is how your utility measures your power usage.

knockout   The removable portion of an electrical box or service panel that, when removed or bent away, allows an electrical cable to pass through.

load   A device that consumes or uses electricity.

main service disconnect   A means to turn off all electrical power to a service panel or fuse box.

modem   A modulator/demodulator. It enables your computer to send digital information over an analog phone line.

National Electrical Code (NEC)   An advisory set of rules for the safe installation of electrical systems.

neutral   A term used to describe the grounded conductor that carries a current back to the source.

neutral bus bar   A metal strip inside a service panel to which all neutral conductors are connected.

nonmetallic cable (NMB)   The most common electrical cable, so called because of its thermoplastic insulation.

ohm   A unit of measurement of electrical resistance.

Ohm's Law   A law of electricity that describes the relationship among voltage, amperes, and resistance; it says that a current, as measured in amperes, is directly proportional to the current's voltage and inversely proportional to the resistance as measured in ohms.

pigtail   A short wire typically connected to another wire with a wire nut to make an easier connection to a device. It originates and stays in the same box.

polarized receptacle   A receptacle with one slot slightly longer than the other. It assures a matching of the related poles of an electrical current.

push-in terminal   A terminal on a receptacle or switch that holds the stripped end of a wire no larger than No.14 by spring pressure.

raceway   A broader term for conduit.

receptacle   A device in your wall that accepts a plug.

rough-in   The initial stage of installing electrical cable and boxes in walls and floors before final connections are made and finish work is done.

sag   A sudden low-voltage condition.

service entrance   The location of an incoming electrical line.

service panel   The circuit breaker panel that connects the house circuits to the incoming power line.

short circuit   An incomplete circuit caused by a hot conductor coming into contact with a neutral, ground, or metal box, causing an immediate fault to the ground. The current flow is not following its intended path.

stud   A vertical, wood framing component of a wall.

surge   A sudden high voltage.

surge suppressor   A device that limits voltage spikes, sags, and surges, limiting damage to any equipment connected to the suppressor.

terminal set screw   A screw that holds the end of a wire tight in a device.

transformer   A device that steps up, or steps down, an AC electrical current from one voltage to another.

travelers   The wires that carry a current between three-way and/or four-way switches.

underground feeder cable (UF)   Electrical cable that can be buried underground.

Underwriters Laboratories (UL)   A national testing agency.

volt   A unit of measurement of electrical pressure or electromotive force.

watt   A unit of electrical power.

zone heating   A system in which individual rooms or areas are heated independently of each other.

Index

burning smells, appliances, 224
buses, 387
buttons, doorbells, 354-355
buying homes, 68
BX cables, 387

C

cable clamps, 387
cable strippers, 103
cabling
  Category 5, 325-327
  coaxial, 324-328
  fiber-optic, 321, 327
  gauges, 157
  home offices, 307-308
  running, 319
  X-10, 328
calculations
  human body resistance, 85
  illumination, 174
  load calculations, 206-208
  Ohm's Law, 30
  watts, 21
cameras, automated, 321
candelas, 174
carbon monoxide detectors, 344-345
Category 5 cable, 325, 327
Cauldwell, Rex, 385
CEC (Canadian Electrical Code), 42-43
ceiling lights, 181
ceiling repairs, 133-134
ceilings, boxes, 216-217
central air, 293-294
central control units, security systems, 341

ceramic boxes, 147
certified journeymen, 7, 388
CFM (cubic feet per minute), 78
CFRA News Radio's Web site, 95
chandeliers, 181
change orders, 132
charges, 29
child-protective caps, 143
children, teaching, 91-93
chimes, doorbells, 353
chucks, drills, 104
CID (Commercial Item Description), surge protectors, 121
circuit breakers, 12, 243
  checking, safety, 6
  double-pole circuit breakers, 257
  fuses, compared, 46-48
  installing, workshops, 303-304
  trips, 47
circuit directories, 242
circuit testers, 45
circuits, 388
  branch circuits, 54
  dedicated circuits, 56, 208, 249
    kitchens, 252-253
  GFCIs
    installing, 58-59
    kitchens, 253
    testing, 60
  lighting circuits, 56
  mapping, 57-58
  overloaded circuits, 225
  overloads, 6
  replacing, 61-62

short circuits, 226-227
small-appliance circuits, kitchens, 253
  testing, 58
  transfers, 313
  types, 56
circular mils, *see* CM (circular mils)
circular saws, 105
clamps, tightening, 192
class 2 wiring, 334
claw hammers, 103
cleanliness, 8
clock receptacles, 142, 255
closets, lighting, 174
CM (circular mils), 23
CO/ALR devices, 227
coaxial cabling, 324-325, 327-328
code
  boxes, 215
  CEC (Canadian Electrical Code), 42-43
  following, 101
  NEC (National Electrical Code), 40-42
  upgrading to, 44
Codecheck.com, 386
color-rendering index, *see* CRI (color-rendering index)
colors, wiring, 19
common screw terminals, 139
Complete Idiot's Guide to Remodeling Your Home, The, 126
computer-controlled systems, 320-322
computers
  surge protectors, 120
  UPS (uninterrupted power supply), 122

F

M

S